Managing Patient Dose in Digital Radiology

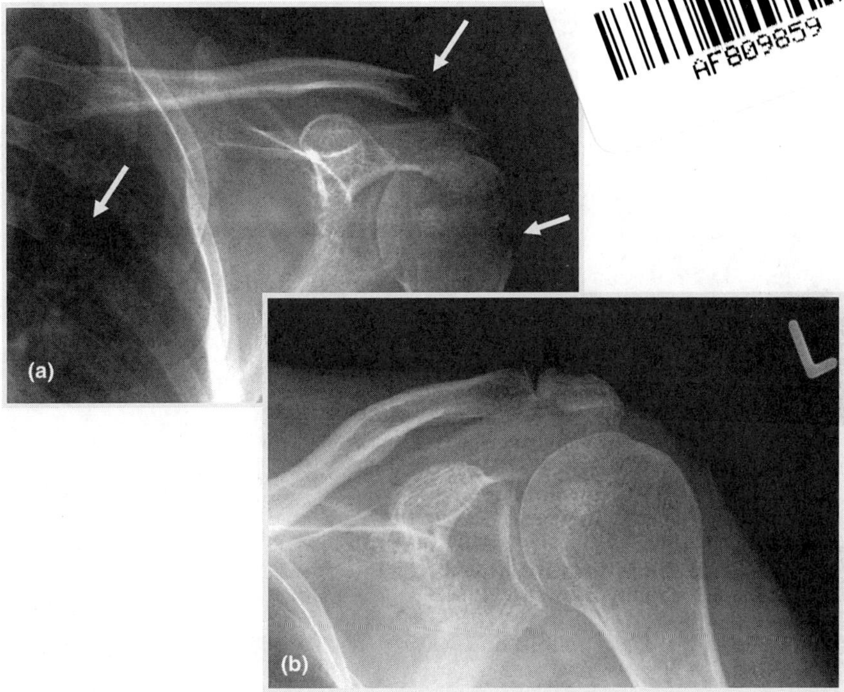

Diagnosis: shoulder pain. (a) Overexposed image with pseudo osteolysis and black hole in the lung. (b) Correct exposure. Agfa: ADC70

A publication from the
International Commission on Radiological Protection

Annals of the ICRP

ICRP PUBLICATION 93

Managing Patient Dose in Digital Radiology

Editor
J. VALENTIN

PUBLISHED FOR
The International Commission on Radiological Protection
by
ELSEVIER

UK	Elsevier Ltd, The Boulevard, Langford Lane, Kidlington, Oxford OX5 1GB, UK
USA	Elsevier Inc., 360 Park Avenue South, New York 10010-1710, USA
JAPAN	Elsevier K.K., 9-15 Higashi-Azabu 1-chome, Minato-ku, Tokyo 106-0044, Japan

Copyright © 2004 ICRP Published by Elsevier Ltd
All rights reserved.

The International Commission on Radiological Protection encourages the publication of translations of this report. No part of this publication may be reproduced, stored in a retrieval system or transmitted in any form or by any means electronic, electrostatic, magnetic tape, mechanical photocopying, recording or otherwise or republished in any form, without permission in writing from the copyright owner.

ISBN 008 044 4695
ISSN 0146-6453

Published quarterly (March issue)

No responsibility is assumed by the Publisher or the ICRP for any injury and/or damage to persons or property as a matter of products liability, negligence, or otherwise, or from any use or operation of any methods, products, instructions, or ideas contained in the material herein. The recommendations and advice of the ICRP reflect understanding and evaluation of the current scientific evidence as given in this report. If and when further relevant information becomes available, the ICRP may review its recommendations. Because of rapid advances in the medical sciences, in particular, diagnoses and administered amounts of radiopharmaceuticals should be independently verified. Although all advertising material is expected to conform to ethical (medical) standards, inclusion in this publication does not constitute a guarantee or endorsement of the quality or value of such product or of the claims made by its manufacturer.

Annals of the ICRP has been published by Pergamon since 1950. In 1991, Pergamon became part of the Elsevier family of imprints. Elsevier carries forward the name and reputation of a publishing house dating from 1580. Elsevier is an integral partner with the scientific, technical and health communities, delivering superior information products and services that foster communication, build insights, and enable individual and collective advancement in scientific research and health care.

Typeset by Scientific Publishing Services, India.
Printed and bound in Great Britain by Polestar Wheatons Ltd, Exeter

CONTENTS

ELSEVIER

ICRP Publication 93

Annals of the ICRP

Managing patient dose in digital radiology

ICRP Publication 93

Approved by the Commission in November 2003

Abstract–Digital techniques have the potential to improve the practice of radiology but they also risk the overuse of radiation. The main advantages of digital imaging, i.e. wide dynamic range, post processing, multiple viewing options, and electronic transfer and archiving possibilities, are clear but overexposures can occur without an adverse impact on image quality. In conventional radiography, excessive exposure produces a 'black' film. In digital systems, good images are obtained for a large range of doses. It is very easy to obtain (and delete) images with digital fluoroscopy systems, and there may be a tendency to obtain more images than necessary.

In digital radiology, higher patient dose usually means improved image quality, so a tendency to use higher patient doses than necessary could occur. Different medical imaging tasks require different levels of image quality, and doses that have no additional benefit for the clinical purpose should be avoided.

Image quality can be compromised by inappropriate levels of data compression and/or postprocessing techniques. All these new challenges should be part of the optimisation process and should be included in clinical and technical protocols.

Local diagnostic reference levels should be re-evaluated for digital imaging, and patient dose parameters should be displayed at the operator console. Frequent patient dose audits should occur when digital techniques are introduced. Training in the management of image quality and patient dose in digital radiology is necessary. Digital radiology will involve new regulations and invoke new challenges for practitioners. As digital images are easier to obtain and transmit, the justification criteria should be reinforced.

Commissioning of digital systems should involve clinical specialists, medical physicists, and radiographers to ensure that imaging capability and radiation dose management are integrated. Quality control requires new procedures and protocols (visualisation, transmission, and archiving of the images).
© 2004 ICRP. Published by Elsevier Ltd. All rights reserved.

Keywords: Managing patient dose in digital radiology; Dose records; Patient dose reduction; Post processing; Image compression

ELSEVIER

ICRP Publication 93

Annals of the ICRP

Guest Editorial

MANAGING PATIENT DOSE IN DIGITAL RADIOLOGY

Digital radiology may represent the greatest technological advance in medical imaging over the last decade, yet many people remain unaware of its existence. With the advent of faster computers, larger storage capabilities, and new x-ray detector systems, film for x-ray imaging is becoming obsolete. An appropriate analogy that is easy for most people to understand is the replacement of typical film cameras with digital cameras. Images can be taken, immediately examined, deleted, corrected, and cropped, and subsequently sent to a network of computers.

In many radiology departments in developed countries, film is simply no longer used in production of x-ray images. The benefits are enormous. The referring physician can often view the requested image on a desktop personal computer, usually accompanied by the interpretation, just minutes after the examination was performed. The images are no longer held in a single location; they can be seen simultaneously by physicians who are kilometres apart. In addition, the patient can have all his or her x rays on a compact disk to take to another physician or hospital.

Digital technology has the potential to reduce patient doses. This is important as medical uses of ionising radiation now represent over 95% of all man-made radiation exposures and is the largest single radiation source after natural background radiation. What then is the problem and why did ICRP Committee 3 request a Task Group to write this document?

While digital techniques have the potential to reduce patient doses, they also have the potential to significantly increase them. Experience has shown that as many radiology departments have transitioned to digital equipment, patient doses have not reduced but have increased measurably. The reasons for this are multiple. Technologists know that an overexposed image can be resolved with the computer, but an underexposed image will need to be repeated. As a result, there is a tendency to give more dose than is necessary. Most systems do not easily track unsatisfactory images that have been deleted from the system, and although the data is present, few systems display meaningful dose or exposure factors for the patient record. Recent studies also show that even if exposures are performed correctly, there is an increase in the number of examinations requested by the referring physicians, probably because of the ease and convenience of getting the images and results.

These are but a few of the issues addressed in this report. It is clear that attention needs to be given to many factors when transitioning to digital systems. A few examples include standardisation of equipment, training, privacy and security concerns, quality control, diagnostic reference levels, archiving, compression algorithms, and referral criteria.

Overall, this is a technology that is advancing rapidly and which will soon affect hundreds of millions of patients. If careful attention is not paid to the radiation protection issues of digital radiology, medical exposure of patients will increase significantly and without concurrent benefit. Conversely, if the radiation protection issues are addressed adequately, medical exposures may decrease without decreasing the diagnostic benefit to the patient.

Fred Mettler
Hans Ringertz
Eliseo Vañó

PREFACE

Over the years, the International Commission on Radiological Protection (ICRP), referred to below as 'the Commission', has issued many reports providing advice on radiological protection and safety in medicine. ICRP *Publication 73* is a general overview of this area. These reports summarise the general principles of radiation protection, and provide advice on the application of these principles to the various uses of ionising radiation in medicine and biomedical research.

Most of these reports are of a general nature, and the Commission wishes to address some specific situations where difficulties have been observed. Reports on such problem areas should be written in a style that is accessible to those who may be directly concerned in their daily work, and every effort should be made to ensure wide circulation of such reports.

A series of such reports was initiated at the Commission's meeting in Oxford, UK in September 1997. On the recommendation of ICRP Committee 3, the Commission has established a number of Task Groups to produce reports on topical issues in medical radiation protection.

Some such reports have already appeared in print as ICRP *Publications 84*, *85*, *86*, and *87*. The present report continues this series of concise and focused documents, and several more advisory reports are being prepared.

The Task Group on Dose Management in Digital Radiology was launched at the Commission's meeting in The Hague, the Netherlands, in September 2001. Its terms of reference were to review methods that are specific to digital radiography that could lead to increased radiation doses. Furthermore, the Task Group was requested to provide recommendations to manufacturers and users that can reduce potential unnecessary doses.

The membership of the Task Group was as follows:

E. Vañó (Chairman)	B. Archer	K. Faulkner
B. Geiger	R. Loose	

The corresponding members were:

B. Bergh	H.P. Busch	J.M. Fernández
R. Gagne	M. Rosenstein	C. Sharp
M. Wucherer		

The membership of Committee 3 during the period of preparation of this report was:

F.A. Mettler, Jr. (Chairman)	J.-M. Cosset	C. Cousins
M.J. Guiberteau	I. Gusev	L.K. Harding (Secretary)
M. Hiraoka	J. Liniecki (Vice-Chairman)	S. Mattsson
P. Ortiz-Lopez	L.V. Pinillos-Ashton	M.M. Rehani
H. Ringertz	M. Rosenstein	C. Sharp
E. Vañó	W. Yin	

This report aims to serve the purposes described above. In order to be as useful as possible for those purposes, its style differs in a few respects from the usual style of the *Annals of the ICRP*.

The report was approved for publication by the Commission in November 2003.

MAIN POINTS

- **In digital radiology, higher patient dose per image usually means improved image quality. This may lead to a tendency to use higher patient doses than necessary with digital systems, and this should be avoided.**
- **Different medical imaging tasks require different levels of image quality. Quality criteria should be established for all medical imaging tasks. The objective is to avoid unnecessary patient doses, i.e. doses that have no additional benefit for the clinical purpose intended.**
- **With digital fluoroscopy systems, it is very easy to obtain (and delete) images, so there may be a tendency to obtain more images than necessary. This would irradiate the patient more than is clinically necessary. Procedure protocols should be agreed to manage this problem.**
- **Local diagnostic reference levels are useful tools to manage patient doses in medical imaging tasks. Diagnostic reference levels for non-digital imaging tasks are not necessarily applicable to specific, similar digital imaging procedures.**
- **Patient dose parameters should be displayed at the operator console (and inside the x-ray room for interventional procedures) to assist radiographers and medical specialists with dose management.**
- **Basic training in the management of image quality and patient dose in digital radiology is necessary for radiologists, medical physicists, and radiographers involved in the use of new techniques.**
- **In the near future, many medical facilities will be filmless, relying more and more on the display and analysis of digital images. This paradigm shift will involve new guidance and regulations, and invoke new challenges for practitioners and radiology employees in the area of quality management.**
- **Justification criteria should be one of the key components considered in the update of a quality assurance programme when a facility converts to digital imaging. The number of examinations could increase because of the relative ease of obtaining and archiving images with digital systems.**

GLOSSARY AND ACRONYMS

Dictionaries of digital terms
http://www4.gartner.com/6_help/glossary/GlossaryR.jsp
http://www.highend3d.com/rosetta/
http://www.learndigital.net/glossary.htm

ACR = American College of Radiology
www.acr.org

ADC = Analogue-to-digital converter
The conversion of analogue signals into digital data – normally for subsequent use in a digital machine.

AEC = Automatic exposure control

Aliasing
The Nyquist condition (see Spatial resolution limit) is not fulfilled if the pixel spacing is too large with respect to the spatial frequency content of the image. The high-frequency components beyond the spatial resolution limit will appear as artificial low-frequency components ('aliasing artefacts') in the image. The practical meaning is that smaller objects will appear artificially enlarged. There is no way to distinguish between a true large detail and an aliased small detail in the final image.

Bit depth
The maximum number of different pixel values is often expressed in powers of 2, which is identical to the number of bits necessary to store the pixel values (binary coding system), e.g. 4096 different pixel values require a bit depth of 12 because $2^{12} = 4096$. Note: The bit depth determines the contrast resolution in connection with the dynamic range.

CCD = Charge-coupled device
A photo-electric device that converts light information into electronic information. CCDs are commonly used in television cameras and image scanners, and consist of an array of sensors that collect and store light as a build-up of electrical charge. The resulting electrical signal can be converted into computer code and reconstructed to form an image (see the ACR standards, available at http://www.acr.org/dyna/?doc = departments/stand_accred/standards/standards.html).

CDRAD phantom
Used for evaluation of digital systems. It is possible to test the image quality and the observer's perception. The phantom consists of a Plexiglass tablet with cylindrical holes.

Characteristic curve (gradation curve)
Describes the mathematical relationship between the absorbed dose at the entrance of the detector and the output signal of the detector. In contrast to the

S-shaped curve of a film-screen system, most digital detectors show a linear relationship.

CR = Computed radiography

A system of creating digital radiographic images that utilises a storage-phosphor plate (instead of film) in a cassette. Once the plate is exposed, a laser beam scans it to produce the digital data that is translated into an image.

CRT = Cathode ray tube

DICOM = Digital Imaging and Communications in Medicine

A set of protocols describing how radiology images are identified and formatted. DICOM is manufacturer-independent and was developed by the American College of Radiology and the National Electronic Manufacturers' Association. The standard emphasises point-to-point connection of digital medical imaging devices. DICOM 3.0 is the current version. http://medical.nema.org/

DIMOND = Dose and Image Quality in Digital Imaging and Interventional Radiology

A European Research Consortium on: Measures for optimising radiological information and dose in digital imaging and interventional radiology. See also http://dbs.cordis.lu/fep/FP5/FP5_PROJl_search.html

Dose

A general term referring to radiation dose (i.e. either radiation dose to a patient or to an imaging device). The word 'dose' is used in this document whenever the context is not specific to a particular dosimetric quantity. When the context is specific, the name or symbol for the quantity is used [e.g. absorbed dose at a point (D), mean absorbed dose in an organ or tissue (D_T), dose-area product (DAP), air kerma (K_a), etc.]. See Appendix B for more complete definitions of the dosimetric quantities used in this document.

Dose index (see Exposure index)

Dynamic range (of an x-ray detector)

The ratio of the maximum and minimum dose that can be accepted by an imaging device without deteriorating or distorting the image. Sometimes this dynamic range is expressed as the ratio of the maximum and minimum pixel values (only valid for a linear relationship between absorbed dose at the entrance of the detector and pixel value). Flat-panel detectors have a dynamic range of 10^4 (from 1 to 10,000), and film-screen systems have a dynamic range of approximately $10^{1.5}$ (from 1 to 30) (Hennigs et al., 2001).

Dynamic range (of an image or object)

The ratio of the maximum and minimum absorbed doses from the projected radiation field at the entrance plane of the x-ray detector. The dynamic range of an image depends mainly on the properties of the anatomical region, but can also be influenced by technique factors (e.g. beam quality).

DQE = Detective quantum efficiency
A physical approach for a general description of imaging performance that compares the signal-to-noise ratio (SNR) in the image to the SNR in the radiation field. A DQE value of unity describes an ideal detector. A DQE value of 50% means that the detector of interest requires twice the absorbed dose to generate an image with the same SNR fidelity as the ideal detector. The DQE is usually given as a function of spatial frequency. The DQE of a real detector also depends on beam quality and absorbed dose level.

EPR = Electronic patient record
Healthcare record stored in electronic format.

Exposure
A general term referring to the act of irradiating a patient for a clinical purpose, or irradiating a device that converts radiation exposure into useful clinical images.

Exposure index
Term used in relation to the absorbed dose to the phosphor plate. After reading the image with the laser system, the signal histogram is computed, and the dose level or exposure index is determined from the pixel values using a logarithmic relationship (Huda et al., 1997).

FDA = Food and Drug Administration (in the USA)

Gradation curve (see Characteristic curve)

Graininess
Related to noise perception. The impression of inhomogeneity of an apparently (or expected) homogeneous area due to noise sources (quantum noise, electronic noise, fixed structures, stochastic anatomical fluctuations, etc.).

Grey scale
Shades of perceived brightness values that are generated by the display device. Each pixel value of the digital image is mapped to a brightness value. For optimum mapping of pixel values into perceived grey shades, the conversion should utilise the concept of the 'grey-scale standard display function' as described in the DICOM standard.

HIS = Hospital information system
A system used to store and retrieve patient information. The HIS is an integrated computer-based system that may include or be linked to laboratory and radiology information systems (denoted as LIS and RIS, respectively).

ICRU = International Commission on Radiation Units and Measurements
www.icru.org

IEC = International Electrotechnical Commission
http://www.iec.ch

Image matrix
Represents the whole set of pixels forming an image. The matrix size is the number of rows and columns in a digital image (e.g. 1024 × 1024). The matrix size of the image cannot be identical to the matrix size of the x-ray detector.

JPEG = Joint Photographic Experts Group
JPEG is a standard for data compression. It offers data compression of between two and 100 times. http://www.highend3d.com/rosetta/?cat = j

Kernel
A parameter used in image processing algorithms to define the impact of the neighbouring pixels on the pixel of interest. The kernel can be interpreted as a calculation scheme usually written in a matrix form, where the pixel currently being processed is in the centre (e.g. a kernel of 3 × 3 specifies that only the information of the adjacent eight pixels is used to recalculate the content of the centre pixel).

LAN = Local area network
A network of a small number of computers whose reach is limited (usually within a building or campus). The computers are linked to allow access and sharing of data and computer resources by users.

LIH = Last image hold
Feature on fluoroscopic systems allowing the continuous display of the last image acquired following termination of any exposure period.

Lossless compression
A method of compressing and storing a digital image in such a fashion that the original image can be completely reconstructed without any data loss.

Modality (or imaging modality)
A generic term used to describe an imaging device such as magnetic resonance imaging, computed tomography, ultrasound, digital radiography, computed radiography, mammography, etc.

MTF = Modulation transfer function
Describes how well the contrast of an object with defined size is preserved in the image. A MTF value of unity means that the contrast has not been degraded at all. A MTF value of zero means that the detail is no longer present in the image. The MTF is usually given as a function of spatial frequency (see ICRU, 1986).

Nyquist (frequency)

The minimum frequency that will faithfully sample an analogue signal. This is always twice the maximum frequency of the signal to be sampled. In practice, significantly higher sampling frequencies are used in order to stay well above the Nyquist frequency and therefore avoid the chance of producing unwanted artefacts.

Nyquist condition (See Spatial resolution limit)

Original image

The image obtained (read-out signal of flat-panel detector or storage-phosphor system) after all device-specific corrections.

PACS = Picture archiving and communication system

A system capable of acquiring, transmitting, storing, retrieving, and displaying digital images and relevant patient data from various imaging sources, and capable of communicating the information over a network.

Pixel (picture element)

A digital image is formed by single pixels. Each pixel represents the image signal at a defined location in the image. As images are usually rectangular in shape, the pixel location is represented by a column and a row co-ordinate.

Pixel aperture

The physical structure in an x-ray detector consisting of all means to convert the local x-ray exposure into an electrical signal is also referred to as a 'pixel'. Depending on the detector technology, these pixels may have different shapes and sizes, or represent a laser spot size as in storage-phosphor systems. The sensitive area forming the image signal is important in the conversion to an electrical signal, and this area is referred to as a pixel aperture.

Pixel spacing

The distance between the centres of the pixel apertures. The pixel spacing determines the spatial resolution limit of the device.

Pixel value

A numerical code assigned to a pixel. It represents, on an arbitrary scale, the absorbed dose in the x-ray detector.

Quantisation

Basically the conversion of the analogue, continuous quantity into a discrete (digital) numerical code. The difference between the discrete and the (real) analogue value is called the quantisation error and limits the achievable low-contrast resolution. This quantisation error should always be significantly lower than the local signal variation due to quantum noise.

RAID = Redundant array of independent disks
A grouping of standard disk drives together with a controller to create storage that acts as one disk to provide performance beyond that available from individual drives.

Raw image (raw = read after write)
Read-out signal of flat-panel detector or storage-phosphor system. The term 'raw data' is often used to emphasise that they are unprocessed.

RIS = Radiology information system
A system that supports the information processing and business requirements of radiology departments and freestanding image centres.

Sampling
The physical process of assigning a single value to a pixel by integrating the signal captured by the pixel aperture. Due to the finite size of the pixel aperture, details that are smaller than the pixel aperture are blurred.

SCAR = Society for Computer Applications in Radiology
http://www.scarnet.org

Sharpness
The visual impression of the quality of reproduction of small details.

Signal normalisation
Scaling of the image signal with regard to the absorbed dose to the detector, by performing the digital processing in a way that the brightness of the displayed image is kept on a defined constant level, independent of the exposure.

SNR = Signal-to-noise ratio
The ratio of noise-to-picture signal information.

Spatial filter
An image processing function for enhancement (or suppression) of anatomical structures depending on the spatial frequency (i.e. the size) of the structure. The most common application is the enhancement of small details by increasing the sharpness impression (so-called edge enhancement). The function is usually controlled by two parameters:
(i) the kernel size in pixels (for controlling the size for enhancement); and (ii) the amount (e.g. as a percentage) (for controlling the enhancement).
Other applications of spatial filters are dynamic range compression or noise suppression (smoothing).

Spatial frequency
Any signal can be composed of a series of harmonic (sine and cosine) waves. An image can be interpreted as a composition of an infinite number of periodic sine and

cosine waves. A short wavelength (equivalent to high spatial frequency) corresponds with small detail, whereas a long wavelength (equivalent to low spatial frequency) corresponds with large objects in the image. The relationship between spatial frequency and detail size is inversely proportional. In order to avoid confusion with the term 'time frequency', 'spatial frequency' is used. A common unit is line pairs per millimetre (lp/mm).

Spatial resolution limit

Following the sampling theorem (Beutel et al., 2000), at least two sampling periods are necessary to detect the frequency of a signal properly. This requires that the sampling frequency be at least twice that of the highest spatial frequency component in the x-ray image (see Nyquist condition). The sampling frequency of the device is determined by the pixel spacing.

Speed classes

The relative 'dose' required to obtain images with appropriate optical density (i.e. blackening). The higher the speed class, the lower the 'dose' required to obtain the given optical density (i.e. the image).

SPIE = The International Society for Optical Engineering
http://www.spie.org/

System gain

The ability of an image intensifier to produce a bright image. With digital systems, gain is one of the factors that need to be optimised in order to produce optimal image quality with an acceptable radiation dose.

Virtual collimation

Allows customisation of the dimensions of the x-ray beam using numerical methods without the need to expose a patient or a phantom. This option is available for some fluoroscopy digital systems.

Zoom in/out

Due to technical limitations, many display devices are not able to show the full image matrix generated by the x-ray detector. Image viewing software supplies zooming algorithms to handle this situation. It should be noted that dropping pixels when fitting the image to the display area can cause aliasing artefacts, and these should be suppressed by image processing (controlled spatial filtering to limit the frequency content of the displayed submatrix). Most workstations provide a specific 'full-resolution' mode that displays exactly one image pixel by one display pixel without any interpolation ('1:1 scaling').

1. INTRODUCTION

Main points

- **Medical imaging is being transformed significantly by the transition from analogue-based film to digital techniques.**
- **The diagnostic information provided by modern digital detectors can be equal or superior to conventional film-screen systems, with comparable patient doses.**
- **Digital imaging has practical technical advantages compared with film techniques, e.g. wide contrast dynamic range, postprocessing functionality, multiple image viewing options, and electronic transfer and archiving possibilities.**
- **With digital systems, an overexposure can occur without an adverse impact on image quality. Overexposure may not be recognised by the radiologist or radiographer. In conventional radiography, excessive exposure produces a 'black' film and inadequate exposure produces a 'white' film, both with reduced contrast. In digital systems, image brightness can be adjusted post processing independent of exposure level.**
- **Digital radiology imaging is currently more expensive than film-based radiology. If managed appropriately, the additional cost can be justified by significantly improved image quality with lower patient doses, improved workflow through shorter examination times with improved patient comfort, reduced archive requirements, and the potential for teleradiology and remote expert consultation.**
- **There are currently two main technical approaches to imaging for digital radiography systems: storage-phosphor plates located in the cassettes of conventional x-ray systems, and the flat-panel detectors of dedicated digital systems.**
- **Digital fluoroscopy systems are usually based on the use of image intensifiers. The output of the image intensifier is projected onto a video camera or charge-coupled device (CCD) chip that converts the signal into a digital image. Some newer systems use a dynamic flat-panel detector instead of the image intensifier and video camera.**
- **A pixel matrix representation of an image is produced by digital radiology (radiography and fluoroscopy). Each pixel is used to store a specific numerical value related to an individual grey level in the image.**

1.1. Purpose of this document

(1) This document is addressed to radiologists, radiographers, medical physicists, and all other interested individuals who are familiar with conventional film-based radiography and fluoroscopy. No specific physics, electronic, or computer knowledge is necessary to understand the basic principles of digital radiography and fluoroscopy or their clinical applications. This document discusses the advantages and disadvantages of digital imaging, aspects of radiation dose, elements of image quality, equipment costs, and new regulations and standards. Other digital techniques, i.e. computed tomography (CT) and digital subtraction angiography (DSA), are beyond the scope of this document.

1.2. Introduction

(2) In the last decade, the replacement of conventional fluoroscopic and radiographic equipment with digital imaging systems has increased rapidly in developed countries. The conversion from conventional to digital image acquisition necessitates additional equipment. Processing, reading, storage, and the distribution of digital images requires computers, digital networks, and laser imagers or picture archiving and communication systems (PACS). The wide dynamic range of digital detectors avoids the requirement for repeat examinations due to over- or underexposed images. However, these dynamic ranges, and post processing, also make it more difficult to recognise over- or underexposure as the images may be of diagnostic quality in either case. The dynamic range of flat-panel detectors can achieve a dose reduction of 30–50% without any decrease in diagnostic performance. Also, the dynamic range provides an opportunity to tailor the patient dose and image quality for a specific medical imaging task, e.g. in paediatric examinations. Additionally, digital techniques allow remote diagnosis and expert consultation by teleradiology.

(3) Digital radiology cannot be recommended for all countries. A minimum level of electronic and mechanical engineering expertise and infrastructure is necessary to support such systems. Film-based conventional radiology, with wet image processing, is simple, stable, and, at present, less expensive.

1.3. History

(4) For more than 100 years, x rays have been used with radiographic films for medical imaging. New developments in computer applications in the last decade have fostered new techniques that have extended the spectrum of radiological imaging significantly. Despite the rapid progress of radiology in the area of magnetic resonance imaging (MRI) and CT, large-projection or plain-film radiography from lung, skeletal, and gastrointestinal examinations still represent up to 80% of all examinations, and are the major part of the daily routine in radiology.

(5) Following the introduction of DSA in combination with fluoroscopic systems at the beginning of the 1980s, digital image intensifiers were introduced into the field of projection radiography. This new method was used primarily for investigations of the gastrointestinal tract and other fluoroscopic examinations. In the following years, the obvious advantages (high image quality, potential for lower patient dose, direct availability of images, simple handling, and the other advantages of digital image information format) led to an almost complete replacement of the film-screen technique in image-intensifier-based systems.

(6) In the mid-1980s, storage-phosphor radiography was introduced as a new digital technique for wall- or table-mounted Bucky systems and bedside examinations. The initial significant technical challenges, financial costs, and limited image quality delayed the wide clinical application of this technique until the beginning of the 1990s. Today, storage-phosphor-based systems play a fundamental role in the field of digital projection radiography.

(7) More recently, flat-panel detectors have been developed that enable direct digital registration of image information at the detector. Since 2000, these detectors have become available commercially and they are slowly being introduced into the clinical routine. Dynamic flat-panel detectors are already replacing image intensifiers in routine clinical practice.

(8) An expanding field of research and clinical application is occurring in digital mammography. The spatial resolution of these systems needs to be higher than that for all other digital systems. The initial flat-panel detectors or storage-phosphor systems developed for mammography produce a spatial resolution of 5–8 line pairs per mm. This resolution is still 50% less than that for conventional film-screen systems.

(9) Although spatial resolution is an important consideration in the evaluation of the image quality of the acquired data, the impact of noise needs attention and is of particular importance when considering noise-limited, low-contrast imaging tasks, e.g. detection of microcalcifications and masses in mammography. Complete evaluation of the image quality of the acquired data requires an appreciation of the imaging task, and the measurement of some fundamental parameters such as grey-scale transfer, resolution, noise, and incident radiation exposure.

1.4. What is digital radiology?

(10) Conventional radiographic images are based on the exposure and processing of films. Hence spatial resolution and grey levels are analogue values. The dynamic range or latitude of film is limited by the maximum optical density that the film can produce. Digital radiography and fluoroscopy, which do not have this limitation, utilise a matrix of discrete numerical values to represent an image.

(11) A matrix is a square or rectangular area divided into rows and columns like a chessboard. The smallest element of a matrix is called a pixel (from picture element). The location of each pixel in a matrix is encoded by its row and column number (x, y). Each pixel of a matrix is used to store a binary number with a range of 8–16 bits (bit is short for binary digit). Eight bits allow the storage of integer values between 0 and 255; 16 bits allow values between 0 and 65,535. Small grey-level ranges of 8 bits are used for ultrasound or film printing with laser imagers, and 10–14 bits are used for digital fluoroscopy or digital radiography. These numerical values are related to the individual grey levels of an image. The contrast in a digital image is represented by the difference in the numerical pixel values in different areas of the image.

1.5. Differences between intrinsic digital modalities and digital radiology

(12) Digital radiography and fluoroscopy are new techniques that are capable of replacing film-based image acquisition. In contrast, intrinsic digital modalities such as CT, MRI, DSA, single photon emission CT (SPECT), and positron emission tomography (PET) have no preceding analogue acquisition. Computers were necessary to obtain digital images from x-ray projection data or radio frequency signals.

Conversion of conventional films into digital images by means of digitisers is pseudo digital radiology.

1.6. Digitised conventional films

(13) Conventional films may be converted into a digital image format and stored electronically by use of a digitiser. Modern digitisers provide sufficient spatial and grey-level resolution for nearly lossless conversion of information. However, the use of digitisers for conversion with digital storage of all images in a radiology department cannot be recommended. High-quality digitising and the creation of a correct digital patient and examination header is time consuming and adds additional expense. Hence, digitising films can only be recommended in selected cases such as the integration of previous digital examinations into a digital archive (i.e. PACS). Other reasons may include the saving of conventional film copies (if a patient leaves the hospital for further external treatment), and digitising of mammograms for post processing. The use of film digitisers has proven useful in mammography. In this particular application, the digitised data serve as a significant adjunct to the physician.

(14) Digitising of films in non-digital departments for creation of teaching files is an important application, as is digitiser-based or video-camera-based image acquisition for teleradiology. Several commercial attempts to introduce film-based radiographic x-ray systems with integrated digitisers have not been successful.

1.7. Plain radiography

(15) Cassette-based film-screen systems have proved to be a stable technique. The advantages are good image quality, a simple acquisition technique, and moderate costs. Disadvantages are a small dynamic range, the lack of post processing, exposure requirements, and the availability of radiographic film at only one location.

(16) The major difference in all digital radiography systems is the technical approach to imaging, i.e. how the physical process of x-ray detection is realised. The following techniques are available:

- storage-phosphor technique;
- selenium-cylinder detector;
- CCDs; and
- flat-panel detectors with direct or indirect conversion.

(17) Storage-phosphor was the first available technique for digital radiography. The storage plates are exposed in cassettes with standard dimensions. Hence, no change of generator, x-ray tube, and wall- or table-mounted Bucky system is necessary. Even bedside examinations and other special projections are possible.

(18) In 1994, the selenium-cylinder technique was the first introduction of a solid-state detector into chest radiography. It has a high efficiency for x-ray detection that is comparable with modern flat-panel detectors. The application is limited to examinations with a vertical patient position (chest, abdomen, and spine).

(19) CCD systems represent a very small share of the market. The image of a luminescent screen is recorded with CCD cameras or devices and converted into digital images.

(20) The latest developments in digital radiography are flat-panel detectors with direct or indirect conversion of x rays into electrical signals. These detectors provide the highest quantum efficiency, excellent image quality, and enable a substantial patient dose reduction [with the appropriate quality assurance (QA) programme]. Portable flat-panel digital radiography units have been evaluated for neonatal imaging; substantial dose reduction, requiring one-quarter of the patient dose compared with conventional radiography, was found, with rapid availability of images for timely clinical decision making (Samei et al., 2003).

1.8. Digital fluoroscopy

(21) Both digital and conventional fluoroscopic systems are based on the use of image intensifiers (except new flat-panel systems). In conventional systems, the output screen of the image intensifier is projected by an optical lens onto a film. In digital systems, the output screen is projected onto a video camera system or a CCD camera. In addition to the digital acquisition, some systems provide additional cassette holders for exposure of conventional films or storage-phosphor plates. Patient dose can be reduced by digital functions, e.g. last image hold (LIH), virtual collimation, pulsed fluoroscopy, image processing, etc.

1.9. Comparison of digital and non-digital techniques

(22) In comparison with conventional radiology, digital radiology needs new components such as a digital detector, computers, networks, laser imagers, workstations, and digital archives. When introducing digital radiography in a radiology department, some aspects of the workflow will need to be changed. Another main difference is the relationship between exposure and presentation of an image. In conventional film-based radiography, there is a well-known relationship between the film exposure and the result of black, grey, or white images. Both under- and overexposures result in changes with reduced contrast. This can be simplified by: 'an excessive exposure is black, an inadequate exposure is white'. In digital radiology, the grey levels of an image are normalised and the brightness of an image is independent of the mean absorbed dose in the imaging device. Normal levels of absorbed dose produce good or normal images, and substantial reductions in absorbed dose increase the pixel noise of an image. Further advantages of digital radiology are the features of brightness adjustment with 'window/level', magnification, post processing, and various types of digital communication. The critical issue is an overexposure that may not be noticed by the radiologist or the radiographer because the images are still of good quality.

1.10. Image quality and exposure levels in digital radiology

(23) Many studies have compared exposure levels and image quality of digital and conventional radiography. They have concluded that the diagnostic information from digital radiography is superior to conventional radiography if the exposure parameters used are equivalent (or for equivalent image quality, digital radiography requires less exposure). All digital techniques are characterised by a much higher dynamic range, preventing over- or underexposure to a large extent. The spatial resolution of all digital systems is about half that of conventional film-screen systems. Despite this, good diagnostic image quality can be secured by postprocessing methods.

1.11. Further aspects of digital radiology

(24) In the last decade, the number of digital radiology systems installed has increased substantially worldwide, replacing many conventional film-based systems, especially in hospitals. Only digital radiology allows the change from wet chemistry to dry laser imagers.

(25) The use of digital radiology equipment requires special training of radiologists, radiographers, and medical physicists. Despite the advantages of digital systems, there are additional risks of undetected patient pathology and occurrence of specific digital artefacts. Some of these artefacts may be misinterpreted as pathology in a patient.

(26) The physical principles of digital image generation and processing require new approaches, such as the use of specific 'digital phantoms' and derived parameters such as 'exposure index'. The postprocessing algorithms of different manufacturers should be harmonised. Access to the original image is necessary to control the physical detector quality.

(27) The decision to select a specific technique depends on the clinical environment. For small hospitals or institutes with low or moderate examination frequencies, storage-phosphor systems are well-suited, flexible, multipurpose tools. Flat-panel detectors are more expensive, so they are more appropriate if there are large numbers of chest examinations or Bucky-table-based examinations.

(28) Industrial advances have already led to cheaper flat-panel detectors with image acquisition rates that can be used for fluoroscopy, and there is the prospect of new storage-phosphor detectors with an aligned crystal structure and, hence, higher quantum efficiency.

(29) Digital radiology is, at present, more expensive than film-based radiology. These higher costs are justified if all of the following advantages of the new digital technique are considered and incorporated into the daily routine: superior image quality or less patient dose; increased speed and improved workflow; teleradiology and expert consultation; reduction of archive space; and shorter patient examination times with increased comfort.

2. PATIENT DOSE AND IMAGE QUALITY IN DIGITAL RADIOLOGY

Main points

- **In digital radiology, higher patient dose per image usually means improved image quality. However, there is a tendency to use higher patient doses than necessary, and this should be avoided.**
- **Different medical imaging tasks require different levels of image quality. Quality criteria should be established for all medical imaging tasks. The objective is to avoid unnecessary patient doses, i.e. doses that have no additional benefit for the clinical purpose intended.**
- **Image quality can be compromised by inappropriate levels of data compression and/ or postprocessing techniques. This is dependent on the modality. Data compression and postprocessing requirements should be defined by the modality and the medical imaging task.**
- **Unprocessed data obtained with digital imaging should be available to the user.**
- **With digital fluoroscopy systems, it is very easy to obtain (and delete) images. There may be a tendency to obtain more images than necessary, which would irradiate the patient more than is clinically necessary. Procedure protocols should be agreed to manage this problem.**
- **Local diagnostic reference levels (DRLs) are useful tools to manage patient doses in medical imaging tasks. Diagnostic reference levels for non-digital imaging tasks are not necessarily applicable to specific, similar digital imaging procedures.**
- **Patient dose parameters should be displayed at the operator console (and inside the x-ray room for interventional procedures) to assist radiographers and medical specialists in dose management.**
- **Basic training in management of image quality and patient dose in digital radiology is necessary for radiologists, medical physicists, and radiographers involved in the use of new techniques.**

2.1. Relationship between image quality (or diagnostic information) and patient dose in digital radiology

(30) In digital radiology, some parameters that usually identify image quality (e.g. noise) correlate well with dose (i.e. absorbed dose in the patient or in an imaging device). For digital detectors, higher doses result in better image quality (e.g. a less 'noisy' image) over a certain range of doses. When the dose is increased, the improvement is in the signal-to-noise ratio (SNR, noise alone also increases with dose). Thus, a tendency to increase doses can occur, especially in those examinations where automatic exposure control (AEC) is not usually available (e.g. bedside applications). Image quality depends on both the spatial resolution and the SNR. The latter depends mainly on the exposure levels. In a digital image, noise blurs spatial resolution and reduces contrast (Herrmann et al., 2002).

(31) Lower doses would clearly have an adverse effect on image quality with regard to noise, and would probably be noticed. However, if the doses drift upwards, the

overexposure will not be noticed because there will be no adverse effect on image quality (FDA, 2001).

2.1.1. Digital radiography

(32) In a digital imaging system, the increase in dynamic range makes it more difficult to recognise over- or underexposure. As such, the initial set-up of such systems may be more problematic. For this reason, it is important to have AEC test protocols, test equipment (including phantoms), and recommendations on changes to the initial set-up of the x-ray generator and QA programmes. The digital imaging system may have a different sensitivity as a function of x-ray energy from the film-screen imaging system being replaced. If this is the case, the AEC should be reprogrammed. This applies either when a current non-digital system is being upgraded with a digital image receptor or when an entire digital imaging system, with x-ray generator, is being installed.

(33) Figs. 2.1 and 2.2 show similar clinical images with very different dose and noise levels. Fig. 2.3 shows images of a test object obtained with two different dose levels.

(34) The wide dynamic range of digital detectors permits a wide range of doses at the detector or at the entrance site on the patient without an adverse effect on image quality. This is not the case with conventional film-screen imaging, where an increase in dose results in an image that is 'too black'. Computed radiography (CR) systems have a linear response to radiation exposure over four orders of magnitude (Huda et al., 1997).

(35) Even for digital detectors, excessive dose may cause signal saturation and a loss of quality (diagnostic information) in some areas of the image (see Fig. 2.4). This

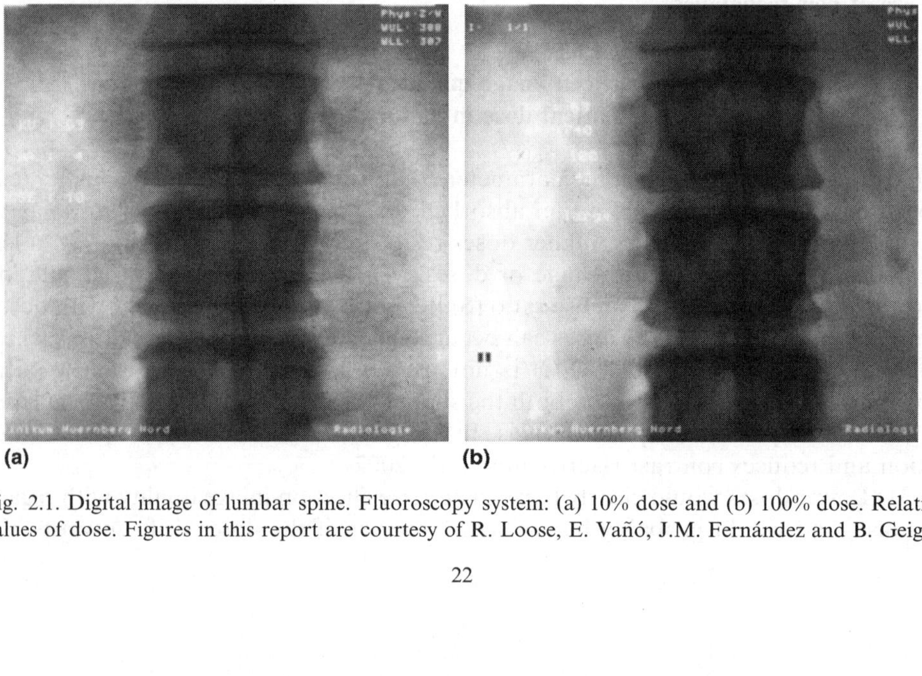

(a) (b)

Fig. 2.1. Digital image of lumbar spine. Fluoroscopy system: (a) 10% dose and (b) 100% dose. Relative values of dose. Figures in this report are courtesy of R. Loose, E. Vañó, J.M. Fernández and B. Geiger.

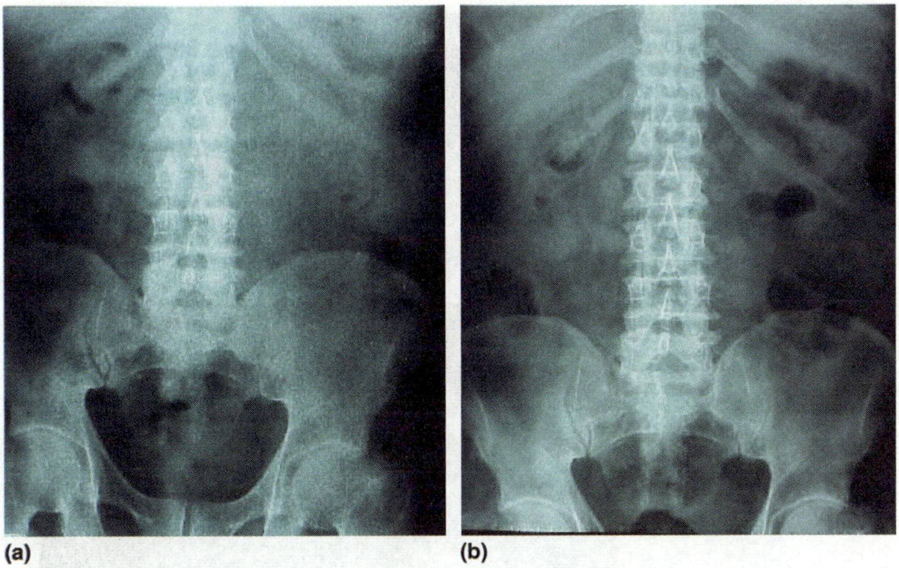

Fig. 2.2. (a) Relative exposure index 1.15, image too noisy. (b) Relative exposure index 1.87, image with sufficient quality.

is easily observable in the lung area of chest images where such excessive doses are used to obtain the image (Fig. 2.5). In such cases, the availability of the original image data could avoid the need for a repeat exposure.

(36) This apparent saturation can occur at doses that are only four- or five-fold greater than the typical dose (e.g. as a result of misuse of the AEC in chest examinations). This is probably due to the factory-set postprocessing criteria. This

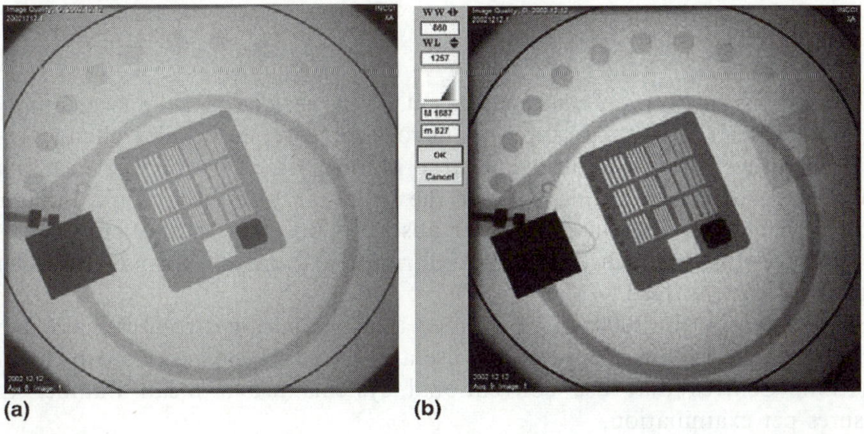

Fig. 2.3. Siemens Axiom Artis. (a) Fluoroscopy low-dose mode, 20 cm PMMA (polymethyl methacrylate), 13 μGy/fr (entrance PMMA). (b) Cine normal mode, 20 cm PMMA, 177 μGy/fr (entrance PMMA).

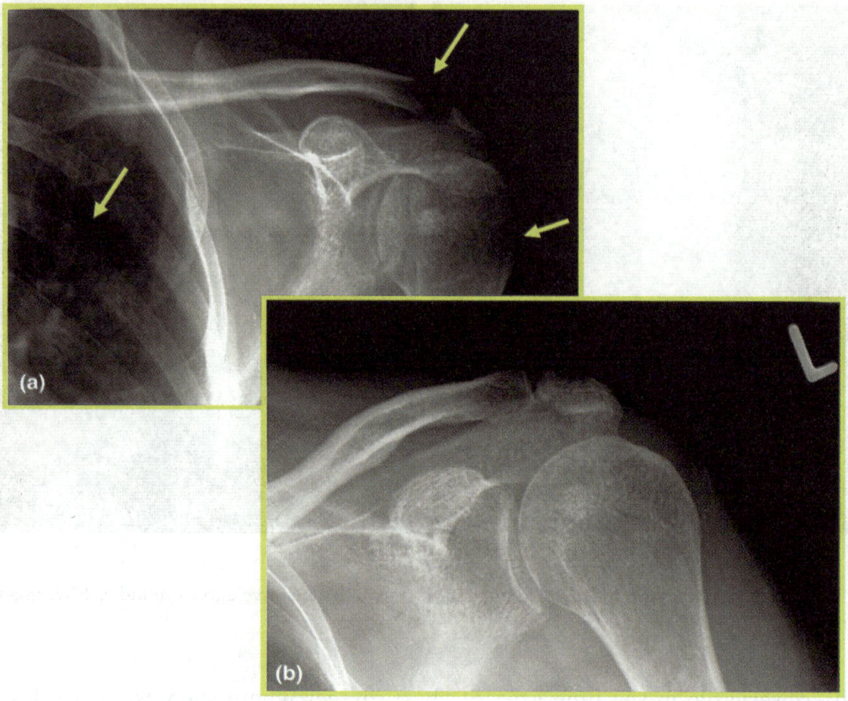

Fig. 2.4. Diagnosis: shoulder pain. (a) Overexposed image with pseudo osteolysis and black hole in the lung. (b) Correct exposure. Phosphor storage, Agfa: ADC70.

emphasises the need to archive original images for a limited period to permit further user-defined post processing, thereby potentially avoiding unnecessary repeat examinations.

2.1.2. Digital fluoroscopy

(37) In digital fluoroscopy examinations, it is very easy to obtain a large number of images because it is not necessary to introduce cassettes or film changers as in conventional systems. In addition, a large number of exposures per examination allows a larger selection of images, with the potential to improve clinical diagnosis. All of these practices result in a higher absorbed dose for the patient. There is a direct relationship between diagnostic information (number of images and image quality) and patient dose.

(38) Axelsson et al. (2000) demonstrated that in upper gastrointestinal examinations, some digital fluoroscopy centres use a mean number of 68 exposures per examination. Centres that use conventional systems use a mean number of 16 exposures per examination.

(39) The immediate availability of the image data and the ease with which new images can be generated can result in an increase in the total number of images, and

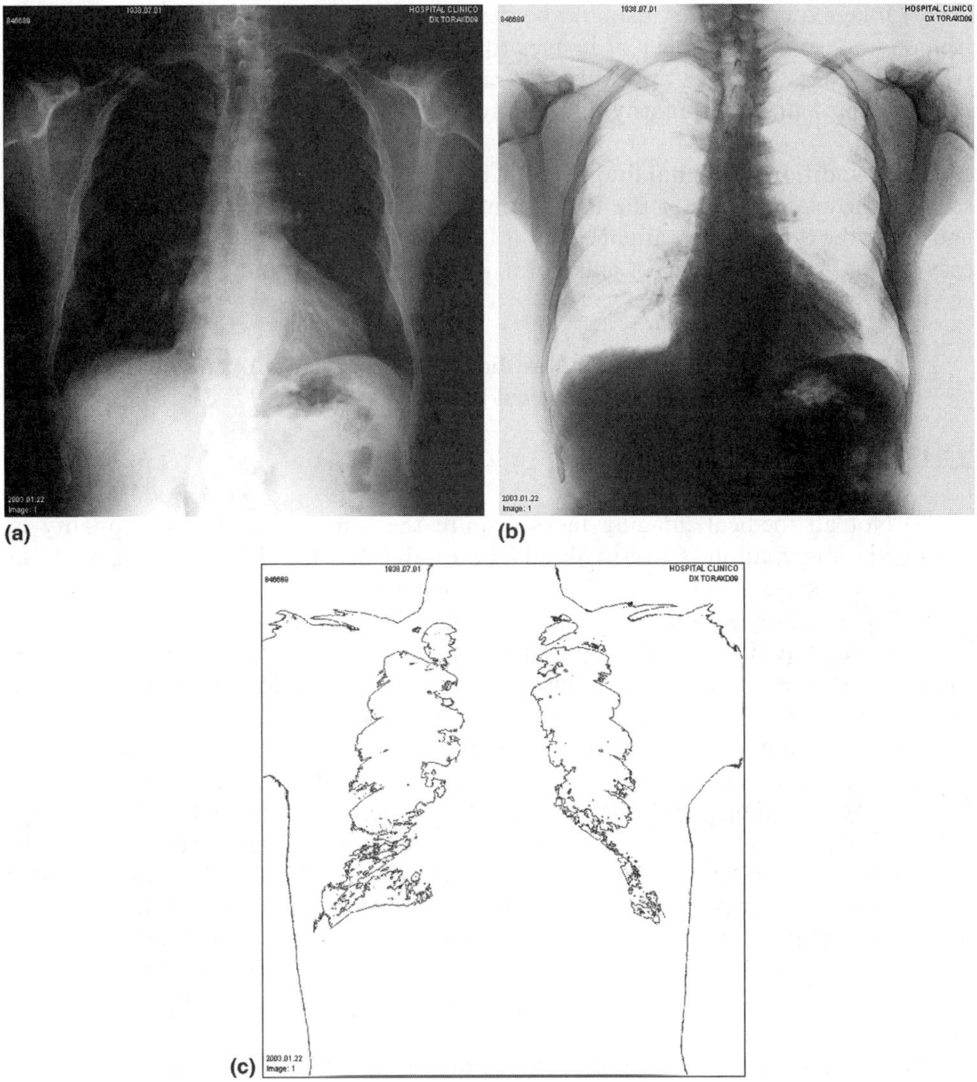

Fig. 2.5. (a) Chest PA (postero anterior) obtained as chest LL (lateral) (flat-panel detector), 125 kV, 6.2 mAs, 0.54 mGy (patient entrance dose four times higher than necessary). AEC centre cell used. Saturated image at the lung area. (b) Same image displayed with inverted grey scale. (c) Same image. Isocontour 99% of pixel content. Some lung areas are saturated without any diagnostic information. The image was repeated.

associated radiation exposure, for a particular radiological examination. This consequence of the ease of use of digital imaging systems needs to be addressed in training materials, on-site training, and continuing education.

(40) Image quality criteria should be developed for digital techniques, and should include a reasonable number of images to be obtained per examination. Diagnostic reference levels would be a useful tool for optimisation of radiation dose if an appropriate patient dose quantity and the number of exposures per examination were recorded.

(41) In addition, in digital fluoroscopy, it is easy to delete unhelpful images before sending the examination to the PACS, but it is difficult to perform a reject analysis and audit the dose to the patient. Even if these images are sent to the PACS, it is not easy to audit the number of unhelpful images routinely and, if done, it is time consuming.

2.2. What image quality (or diagnostic information) is needed for a medical imaging task?

2.2.1. Projection (plain) radiography

(42) Not all medical imaging tasks require the same level of image quality or diagnostic information. Criteria should be established for different imaging tasks to avoid excessive doses where there is no clear net benefit in the diagnosis or the image quality. For instance, the follow-up of a fracture does not require the same level of image quality as that required for its diagnosis. For tasks such as routine follow-up studies, assessment of instability, or orthopaedic measurements, a radiation dose reduction of up to 75% with digital techniques in comparison with conventional film-screen techniques (speed of 400) has been demonstrated (Strotzer et al., 1998).

(43) Some existing guidelines for quality criteria in conventional radiology (CEC, 1996) could be applied to digital studies initially, although some of the criteria may not be relevant for digital examinations (especially the criteria related to the level of contrast in the images). Some indication of the speed and the film-screen combination is given in these guidelines in addition to anatomical criteria. These criteria could also be used for the initial selection of patient dose in digital systems (but the criteria should subsequently be optimised for digital examinations).

(44) A three-level classification of image quality (i.e. high, medium, and low) has been developed in the DIMOND project (see Glossary and Section 3.7) for image quality and patient dose management in digital radiology (Busch, 2003; Busch et al., 2003). Relative 'dose' values (equivalent to speed classes 400, 800, and 1600 in film-screen imaging) are assigned to levels of image quality for the specific imaging method. The higher the numerical value of the speed class, the lower the patient dose.

(45) Examples of clinical imaging tasks requiring high, medium, or low image quality (Busch, 2003) are listed in Table 2.1.

(46) An example of image quality classification results and 'dose' assignments from a comparative study between flat-panel, storage-phosphor (i.e. CR), and film-screen imaging methods is shown in Table 2.2.

Table 2.1. Clinical imaging tasks requiring high, medium, or low image quality (Busch, 2003)

Clinical problem	Image quality	Comment
Primary bone tumour	High	Image may characterise the lesion
Chronic back pain with no pointers to infection or neoplasm	Medium	Degenerative changes are common and non-specific. Mainly used for younger patients (e.g. less than 20 years of age, spondylolisthesis etc.) or older patients (e.g. more than 55 years of age)
Pneumonia adults: follow-up	Low	To confirm clearing, etc. Not useful to re-examine patient at less than 10-day intervals as clearing can be slow (especially in the elderly)

Table 2.2. Levels of image quality (figure in parentheses is the speed class)

High	Medium	Low
Flat-panel (400)	Flat-panel (800)	Flat-panel (1600)
Storage-phosphor (200/400)	Storage-phosphor (400)	Storage-phosphor (800)
Film-screen (200)	Film-screen (400)	Film-screen (800)

2.2.2. Digital fluoroscopy

(47) During complex examinations utilising both fluoroscopy and acquired images, the requirements for image quality are not the same throughout the procedure. Various fluoroscopy modes and different levels of magnification are available. Each mode delivers a different level of patient dose. Magnification or high-dose modes may only be necessary during part of the procedure (e.g. quantification of a stenosis).

(48) The user can select a number of options for image quality and dose per image during the acquisition of a series of images. Dose per image can be varied by a factor of 10 (depending on the image quality required). Consequently, protocols and guidelines for the management of radiation doses are very important. The protocols should include the requirement for a programmed series of images to be terminated once the required diagnostic information has been obtained.

(49) Diagnostic reference levels have been recommended as a useful and practical tool to manage patient doses (ICRP, 1996, 2001b). In the case of fluoroscopy examinations, it is helpful to consider the number of images and the fluoroscopy time in establishing DRLs (Vañó and Gonzalez, 2001; Vañó et al., 2001).

2.2.3. Procedures to evaluate image quality

(50) The first step in the evaluation of image quality is with test phantoms and associated software. The main physical parameters for the evaluation of image quality (spatial resolution, threshold contrast, noise, detective quantum efficiency, etc.) may be assessed with special test phantoms.

(51) The evaluation of physical parameters is not enough to optimise the conditions for obtaining clinical images. Receiver operating characteristic studies are a more robust way to achieve optimum conditions (Launders et al., 2001).

(52) Clinical image quality is usually evaluated using anatomical scoring criteria utilising experienced radiologists. This is necessary as there are several factors that contribute to image quality in digital radiology (e.g. type of monitor, skills in image manipulation, etc.). Consequently, the evaluation of image quality should be undertaken before new systems and/or software are used in clinical practice.

2.3. Actions that affect patient dose

(53) Table 2.3 summarises actions that can influence image quality and patient dose, and includes an assessment of the impact of each action. Following the table are specific examples of dose reductions or improvements in image quality.

(54) The following are specific examples of dose reduction or image improvement.

- Strotzer et al. (2002) demonstrated a dose reduction of 33% in chest radiography using a flat-panel detector.
- Fink et al. (2002) reported a dose reduction of 50% for chest imaging with a flat-panel detector in comparison with a conventional film-screen imaging system with a speed of 200.
- Herrmann et al. (2002) demonstrated a dose reduction of 50% for chest radiography using a flat-panel detector.
- Hennigs et al. (2001) compared a digital chest radiography system that utilises a large-area silicon flat-panel detector with conventional radiography for visualising anatomic regions of the chest. For the posterior–anterior (PA) view, the digital system performed significantly better for the overall appearance and for all anatomic regions except for the peripheral pulmonary vasculature and hilum where no significant difference was found. For lateral digital images, the trachea, costodiaphragmatic recess, and hilum anatomical regions were rated significantly worse. The retrosternal and retrocardiac lung regions were rated significantly better.

(55) Often, it is not possible to directly compare values for dose reduction because of differences in detector technologies and reference systems between facilities.

2.4. Practical advice for managing patient dose and image quality

2.4.1. Initial setting

(56) When introducing a new digital system into clinical practice, the system should be set up to achieve the best balance between image quality and patient dose. Launders et al. (2001) published methodology examples for selecting the optimum tube potential range for chest examinations. Some amorphous selenium-based x-ray systems are optimised for chest examinations at a lower tube potential range (90–110) than the normally recommended tube potential (150) for other types of system.

Table 2.3. Actions that can increase or reduce patient dose

Action	Influence on patient dose	Influence on image quality or diagnostic information
General (for projection radiography and fluoroscopy)		
Reduction of noise perception in the image (i.e. the perception of the signal-to-noise ratio)	Increase	Improvement
Significant reduction of noise (with saturation of the detector in some areas) (e.g. for the lung in chest images)	Increase	Deterioration, retakes
Deletion of image files at the viewing station or workstation of apparently non-useful images[a]	Increase	Possible loss of some useful information Difficult to control repeated exposures
Allowing poor conditions in the use of the visualisation monitor (Kunz and Busch, 2000) (e.g. insufficient brightness or contrast, poor spatial resolution, etc.)	Increase	Loss of information
Improvement in use of the capabilities of the workstation to visualise images (window and level, inversion, magnification, etc.)	Decrease	Allows more information to be obtained from the same image
Existence of postprocessing problems, problems in the digitiser, printer, local hard disk (see Figs. 2.6–2.13), faults in electrical power supply, problems in the network, etc. during the archiving of images	Increase	Occasional loss of images or retakes
Loss of images in the network or the PACS due to improper identification or other reasons (Smith and Berlin, 2001)	Increase	Retakes
Existence of false lesions or pathologies due to artefacts introduced by incorrect digital post processing (Hennigs et al., 2001)[b] (see Fig. 2.4)	No effect	Loss of information and need for retakes
Images stored in the PACS cannot be (sometimes) post processed	Increase	Does not permit re-analysis of images Possible retakes
Use of different post processing (could sometimes avoid repetitions) (Herrmann et al., 2002)	Decrease by avoiding retakes	Improvement
Allowing easy access to the PACS and teleradiology to look at previous images (Bergh et al., 2000)	Decrease	Improvement
Use of digital radiology to obtain an unjustified increase in the number of procedures (Reiner et al., 2000)	Increase	Information not always necessary
Implement dose indication on the console of the x-ray system (Faulkner, 2001; Zoetelief and Faulkner, 2001)	Decrease[c]	No effect
Specific for projection radiography		
Permitting incorrect calibration or misuse or lack of use of automatic exposure system	Increase	Degradation, retakes
Use deteriorated storage-phosphor plates (see Figs. 2.14–2.17)	Increase due to retakes	Loss of quality, retakes

Table 2.3. (*continued*)

Action	Influence on patient dose	Influence on image quality or diagnostic information
Reduce the number of images per procedure (e.g. avoid the lumbosacral spine image)[d]	Decrease	Remains unchanged
Use appropriate tube potential (see Launders et al., 2001). In general, establish correct radiographic techniques for digital systems	Decrease	Slight deterioration or improvement
Availability of a workstation for post processing (also for radiographers) to avoid some retakes	Decrease	Improvement
Specific for digital fluoroscopy		
Increase the number of images per examination with digital fluoroscopy (Partan et al., 2001)	Increase	Improvement
Use magnification (the use of small field sizes with the image intensifier or flat-panel detector) to improve spatial resolution	Can increase skin dose	Improvement
Use of high-dose fluoroscopy or high-dose mode in digital acquisition	Increase	Improvement
Use of digital serial radiographs (26 frames/s) instead of fluoroscopy	Could increase	Possible improvement
Use of virtual collimation	Decrease	No effect
Use of pulsed fluoroscopy	Can decrease	Sometimes slight deterioration

PACS, picture archiving and communication system.

[a] If a dose register exists on the x-ray system or in the radiology information system, one would know which images had been deleted and would have some basic dosimetric information.

[b] The use of post processing to enhance visibility of some structures may lead to an increase in false-positive diagnoses (Hennigs et al., 2001).

[c] Can avoid repetitions and helps to optimise radiographic techniques.

[d] Using adequate post processing, it is possible to obtain more information from previous images avoiding extra projections that could be usual in film-screen radiology (e.g. lumbosacral junction projection in lumbar spine examinations) (Weatherburn and Bryan, 1999).

(57) For CR, manufacturers offer exposure indices (see 'dose index' in Glossary) as a safeguard against overexposure, but the basis for the recommended values is unclear. Peters and Brennan (2002) established an optimum exposure index to be used as a guideline for a specific CR system to minimise radiation exposures for computed mobile chest radiography, and compared this with the manufacturer's guidelines and current practice. Rill et al. (2003) compared results for three commercially available CR systems, with the goal of implementing the findings clinically.

2.4.2. Operation of digital systems

(58) The number of exposures (i.e. acquired images) per examination and the level of image quality should be limited to the number required for the clinical imaging task (and no more). The advantage of post processing is that more information can be extracted from the modification of window and level, use of numerical magnifi-

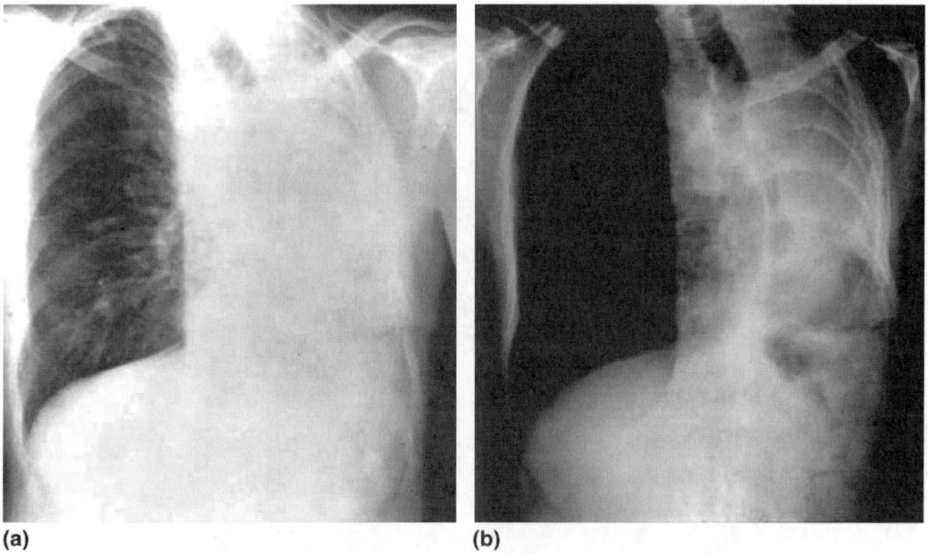

Fig. 2.6. Two postprocessings of a patient after pneumonectomy from raw image. (a) Lung tissue, (b) mediastinum.

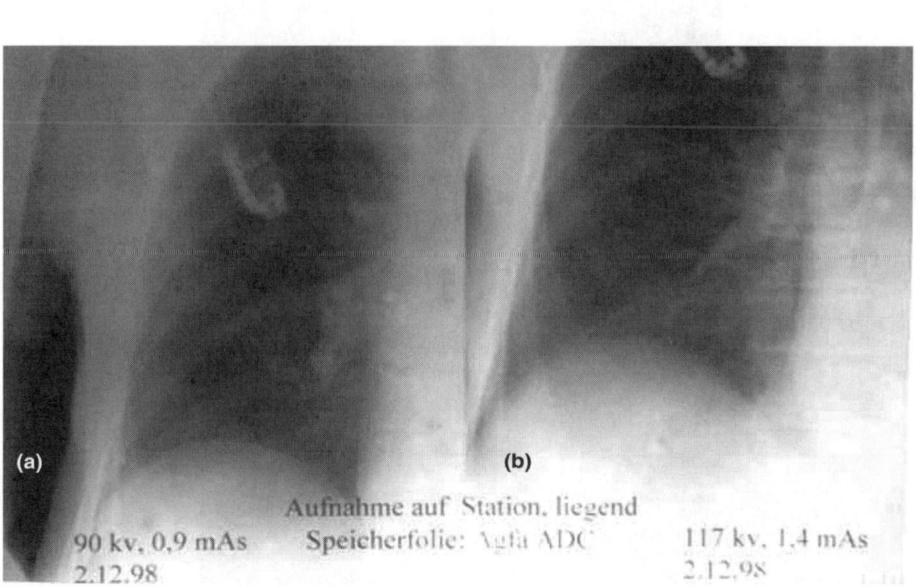

Fig. 2.7. Bedside digital radiography (ADC70) (a) without grid, 90 kV, 0.9 mAs and special post-processing, and (b) with grid, 117 kV, 1.4 mAs and standard postprocessing. The relative dose reduction is more than 50%.

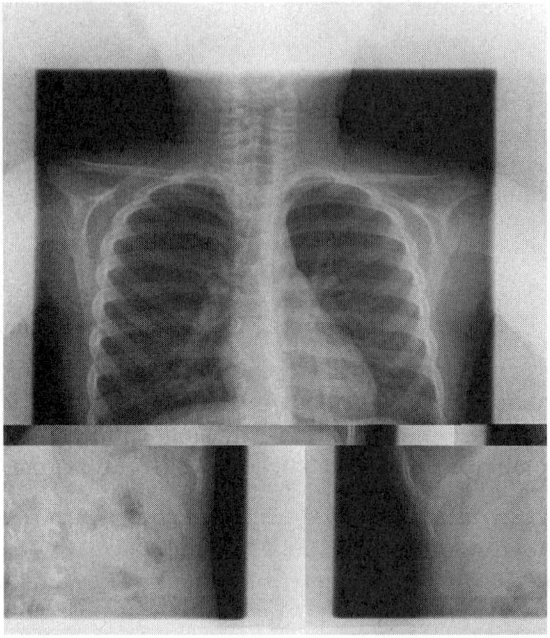

Fig. 2.8. Hard disk problem during archiving in the PACS. Several parts of images are merged and the resulting clinical image is incomplete.

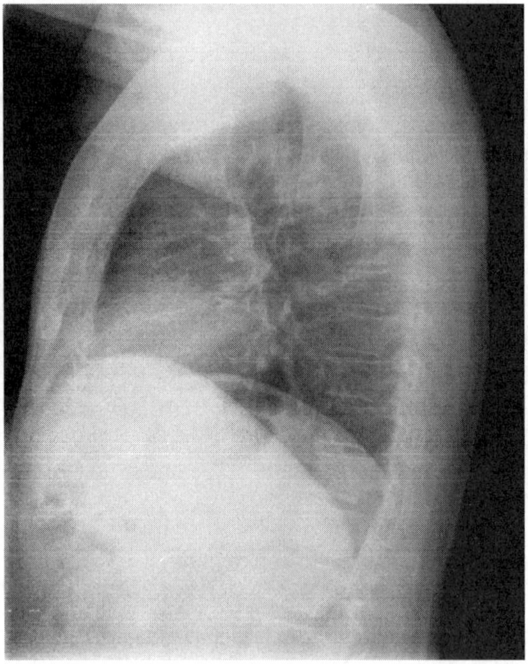

Fig. 2.9. Digitiser problem. Artefacts in the horizontal direction.

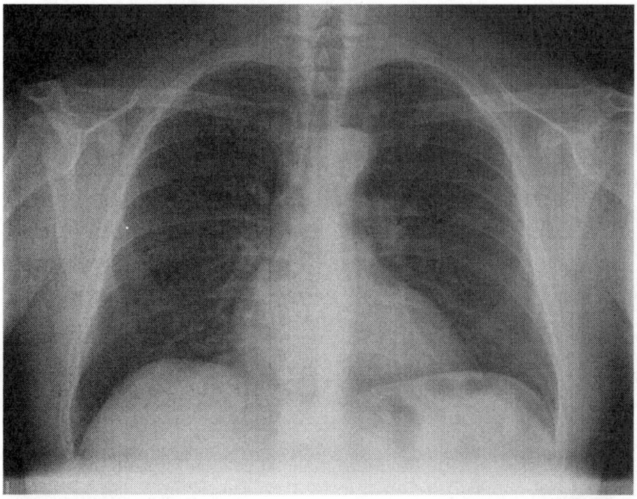

Fig. 2.10. Digitiser problem. Artefacts in the vertical direction.

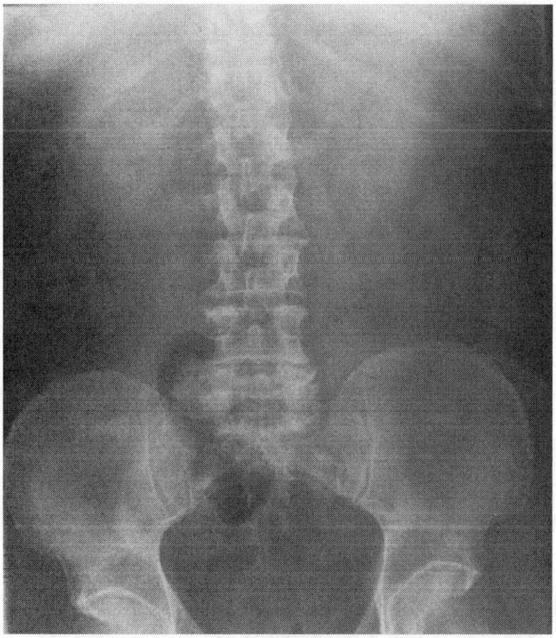

Fig. 2.11. Digitiser problem. Artefacts in the horizontal direction.

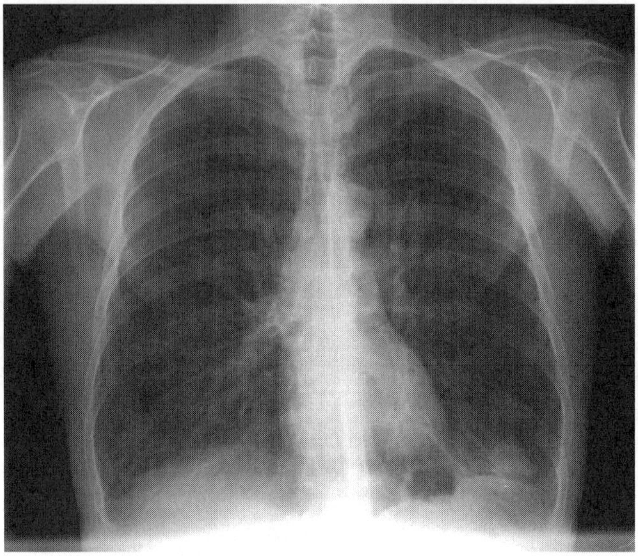

Fig. 2.12. Digitiser problem. Saturation in the lower part of the image.

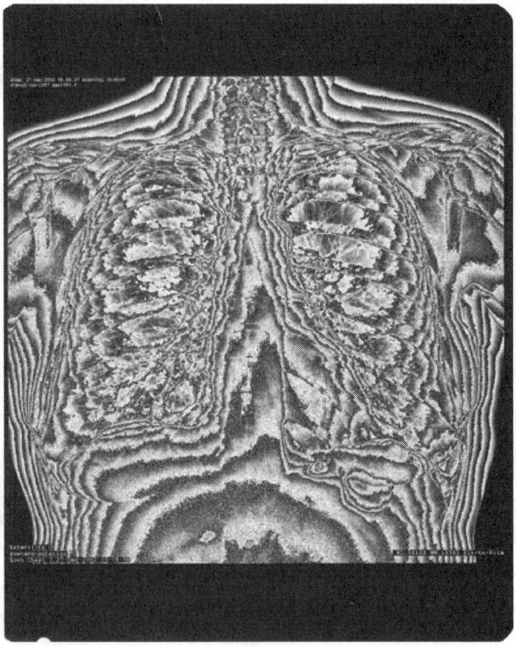

Fig. 2.13. Chest image with printing error.

Fig. 2.14. Phosphor plate artefact.

Fig. 2.15. Phosphor plate misused. Impossible to clean.

cation, grey-scale inversion, etc. Post processing may avoid the extra projections frequently required in some film-screen radiology.

2.4.3. Displaying and taking advantage of dose-related parameters

(59) If patient dose parameters are displayed at the operator console (and inside the x-ray room for interventional procedures), this will assist radiographers, radiologists, and medical physics specialists in the management of patient doses to agreed protocols.

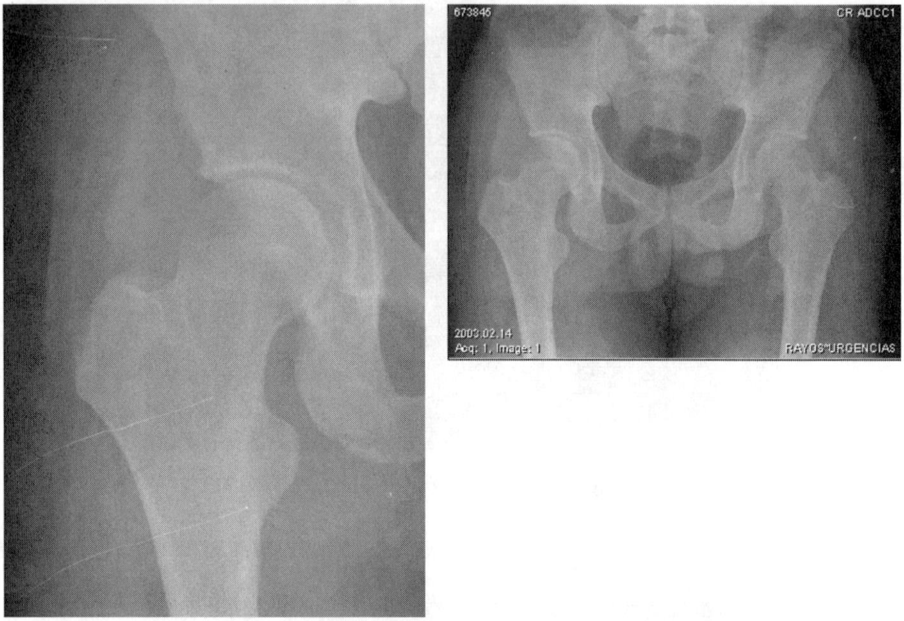

Fig. 2.16. Phosphor plate deteriorated.

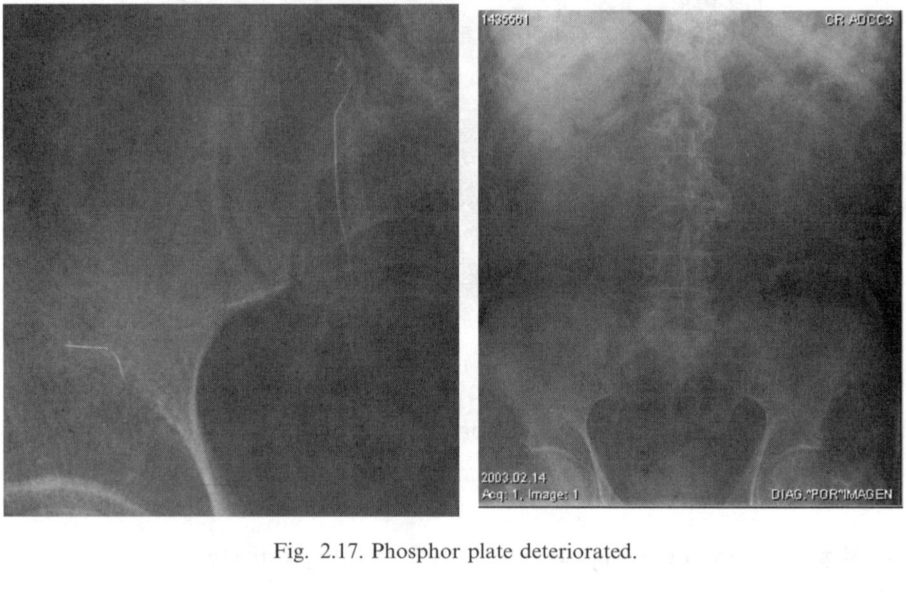

Fig. 2.17. Phosphor plate deteriorated.

(60) Some digital systems offer an exposure index (see Glossary) (e.g. displayed as a colour code in the previsualisation monitor). This exposure index indicates to the operator whether the absorbed dose to the detector is in the normal range (e.g. green or blue colour) or whether it is too high (e.g. red colour). This is a useful training and

Table 2.4. Present and desired future situations for the different digital technologies for data on patient doses

Digital technology	Available now	Desired in the future
Computed radiography	Dose or exposure index	Link with radiographic technique, patient dose estimation, and patient data Archive in the RIS
Digital radiography	Radiographic technique, patient data, and patient dose estimation	Automatic extraction of information from the DICOM header Archive in the RIS
Fluoroscopy	Radiographic technique, radiation field geometry, and dose parameters per series	Fluoroscopy information. On-line skin dose maps and automatic extraction of information from the DICOM header Archive in the RIS

RIS, radiology information system; DICOM, Digital Imaging and Communications in Medicine.

quality control (QC) tool, but the exposure index is not directly related to the patient dose. Currently, different manufacturers calculate the exposure index in various ways, and it would be beneficial if a standardised approach could be agreed.

(61) Real-time collection of dose data, i.e. an easily visible indication on the control panel, would allow comparison with DRLs, facilitate dose management, and help prevent excessive patient doses (Vañó et al., 2002). It would also facilitate the automatic compilation of any individual patient's dose records and provide data for retrospective audits of patient doses in a particular facility (Mosser et al., 2002).

(62) Table 2.4 describes the present situation and the desired future situation for the different digital technologies for data on patient doses.

2.4.4. Special considerations for digital fluoroscopy systems

(63) For digital fluoroscopy systems, several field sizes at the image detector and several operational modes are available for fluoroscopy (usually low-, medium-, and high-dose-rate modes) and image acquisition (sometimes up to four levels of image quality and dose). Operational modes are sometimes combined with factory- (or commissioning-) set postprocessing algorithms as well as with the geometrical set-up (C arm orientations, collimation, focus skin and focus image detector distances, etc.). In addition, service engineers can adjust these operational modes to different users, combining several types of filters, pulse rate modes, dose per pulse, etc. Radiologists should be informed about these settings to allow them to optimise the procedures.

(64) Techniques to reduce patient dose include: (i) collimation prior to irradiation (virtual collimation); (ii) high beam filtration with copper or tantalum filters; (iii) semitransparent filters to compensate for differential absorption in particular examination areas; and (iv) region-of-interest techniques for fluoroscopy.

(65) Pulsed fluoroscopy has the potential to reduce patient dose. However, it requires a service engineer to tailor the dose per pulse, pulse width, and pulses per second, and configure the use of prior data information to fill interpulse gaps. This

will offer adequate image quality for different examinations and, importantly, also reduce dose. In some systems, the available exposure modes can be programmed as required for the type of examination.

(66) With all the possible options in operational conditions, it is essential to obtain the assistance of medical physicists and service technicians to select the optimal operational conditions at a facility.

(67) Some digital systems will be used for interventional procedures. Deterministic effects will be more important in such procedures, e.g. skin injuries. Guidance on managing patient skin doses is not discussed in detail in this report. *Publication 85* (ICRP, 2001b) provides advice on interventional procedures using fluoroscopy.

(68) Radiation protection for the staff is also important during 'in-room' operation of digital fluoroscopy systems. Scattered radiation dose rates are higher during image acquisition than during fluoroscopy. If there is a tendency for the number of acquired images to increase, staff doses are also likely to increase.

2.4.5. Impact of image compression

(69) Image compression can be an important aspect influencing the image quality of stored images in the PACS, and modifying the time necessary to have the images available (transmission speed through networks) (Bergh and Felix, 1999). Loss of image quality can result if the level of compression is too high. Consequently, some examinations may have to be repeated, so the patient will receive an additional dose of radiation (Tuinenburg et al., 2000).

(70) The level of compression compatible with necessary image quality is very dependent on the modality (CR, flat-panel, digital fluoroscopy, etc.) and the type of examination. Therefore, the appropriate level of compression should be established for each modality at each centre, taking into account the centre's network infrastructure and the hardware available across the centre.

(71) Slone et al. (2003) studied whether digital chest images could be compressed for primary interpretation without perceived loss of fidelity (i.e. below the visually lossless threshold) using a transilluminated film or cathode ray tube (CRT) display. They concluded that at the reading distance for primary interpretation, full-size digital chest radiographs, using Joint Photographic Experts Group (JPEG) compression of 10:1 or 20:1 and reconstructed, are visually lossless with film or CRT display. Images compressed to 10:1 remain visually lossless on closer inspection.

(72) If images are to be distributed to different hospital services to be reviewed on conventional computer monitors, a high level of compression could be used. This will increase the transmission speed substantially. Highly compressed images should only be distributed to other hospitals for the purposes of a second opinion or for information, and not for diagnosis. Special QA programmes need to be implemented if this is intended. It should be appreciated that if the level of compression is too high (e.g. to see images on a computer monitor), images may not retain all the original information.

2.4.6. Diagnostic reference levels

(73) In digital radiology, it is relatively easy to improve image quality by increasing the patient dose. Consequently, there can be a tendency for an upward drift in patient doses to levels higher than necessary. For this reason, the evaluation of patient doses as part of QA programmes should be undertaken more frequently than in film-screen radiology.

(74) Re-evaluation of local DRLs should be undertaken when digital techniques replace film-screen techniques, as DRLs for non-digital imaging tasks are not necessarily applicable to similar digital imaging procedures.

2.4.7. Checklist of practical advice

(75) A useful checklist of practical advice for managing patient doses in digital radiology is given below.

- When introducing a new digital system into clinical practice, the system should be set up to achieve the best balance between image quality and patient dose.
- Avoid deleting non-useful images at the workstation and carry out a statistical rejection rate analysis periodically.
- Be familiar with your workstation capabilities (postprocessing capabilities, options in the monitor to visualise the images, etc.).
- Identify all the images correctly to avoid their loss in the PACS.
- Ask for an AEC calibration that is appropriate for the sensitivity range of the digital system and the selected post processing.
- Avoid manual exposures if the AEC is usable. However, ensure it is used correctly, as not all errors are correctable by post processing (e.g. a possible error can be caused by obtaining PA chest images using the central sensor instead of the lateral sensors – this may require a further exposure).
- Control the number of images per examination to maintain it at a number similar to (or lower than) that for conventional radiology.
- Avoid acquiring a digital series of images instead of using fluoroscopy (the patient dose will usually be much higher for the digital series).
- Make a workstation available for radiographers to post process images in order to avoid retakes.
- Pay attention to the dose indication on the panel of the x-ray system or in the in-room monitors, and utilise the information to manage patient doses.
- Establish easy access to the PACS to review previous images in order to avoid retakes.
- Pay attention to proper collimation for the desired anatomical area. Once the image is acquired, numerical methods (i.e. software) may automatically crop part of the image, and when the image is received for reading, the radiologist will not be aware that a larger anatomical area than necessary was irradiated.
- Select the correct preprogrammed technique, e.g. using an abdominal technique (70–80 kV) for chest imaging (120–130 kV) will result in a higher entrance surface dose.

2.5. Training needs for radiologists and radiographers

(76) Specific training in digital radiology is necessary for radiologists, medical physicists, and radiographers involved in the use of digital techniques. This training should include basic aspects of radiation protection for patients and staff, details of the operation of the installed x-ray systems, use of visualisation units, postprocessing capabilities, and the operation of the PACS, if one is installed (Bergh et al., 2000; Busch, 1997).

(77) The European Consortium DIMOND (Dose and Image Quality in Digital Imaging and Interventional Radiology) surveyed radiologists, medical physicists, and radiographers in seven European hospitals about available training resources and relevant training issues. One of the main findings of the survey was a lack of educational resources offered by national and international institutions for digital radiology. The survey also indicated that in the process of transition to digital radiology, between 30% and 35% of professionals were trained on the job and another 23–28% were trained on the job with assistance from the manufacturers of the digital equipment. DIMOND is producing a guideline with specific training objectives for computed and digital radiography, digital fluoroscopy, digital mammography, and aspects concerning digital visualisation and reporting. Further to the survey, suggested hours of training have been proposed (Peer et al., 2003).

(78) An outline of training topics for digital radiology is given below, and a more detailed list of the contents of a training programme is presented in Appendix C.

- X-ray systems for digital imaging.
- Image acquisition.
- Image post processing.
- Image transfer.
- Image analysis.
- QC.
- Aspects in digital visualisation and reporting.
- Workstations.
- Image presentation.
- Image transfer and archiving.
- Patient dose management.

(79) Digital fluoroscopy systems are sometimes used by doctors who are not trained in radiation protection, and this could influence the management of patient dose (Archer and Wagner, 2000). Training for these individuals should include the basic aspects of radiation protection for patients and staff.

3. REGULATORY ASPECTS AND QUALITY MANAGEMENT

Main points

- **In the near future, many medical facilities will be filmless, relying more and more on the display and analysis of digital images. This paradigm shift will involve new guidance and regulations, and invoke new challenges for practitioners and radiology employees in the area of quality management.**
- **Commissioning of digital systems, or introduction of new techniques, should involve clinical specialists, medical physicists, and radiographers. The objective of this collaboration is to ensure that imaging capability and radiation dose management are integrated to achieve acceptable clinical image production using appropriate patient doses. The production of agreed procedure protocols (including justification and optimisation principles) will be an essential part of the process.**
- **Justification criteria should be one of the key components considered in the update of a QA programme when a facility converts to digital imaging. An increase in the number of examinations could occur because of the relative ease of obtaining and archiving images with digital systems.**
- **Once digital systems are in use, comprehensive QC programmes are required to ensure that image quality and patient dose management are maintained. Quality control programmes should detect any significant changes in image quality or patient dose management, and therefore prevent upward 'drift' in doses without additional clinical benefit.**
- **Diagnostic reference levels in digital radiology can act as effective comparators for patient dose management, helping to avoid unnecessarily high patient doses.**

3.1. Radiation protection principles

3.1.1. Justification

(80) As with a conventional x-ray examination, the decision to perform a digital radiology examination needs to be justified. The concept and application of the fundamental principles of justification of medical radiation exposures have been described elsewhere (IAEA, 1996; ICRP, 1977, 1991).

(81) Medical practitioners and radiologists are frequently required to make decisions on the use of radiological procedures. In doing so, they should:

- evaluate the potential value of the procedure in relation to patient management, and evaluate the risks and benefits arising from the procedure (e.g. is there a good chance of obtaining the necessary clinical information and is it likely to influence the management of the patient's illness?);
- determine the best procedure to aid in the diagnosis; and
- consider the availability of results from previous examinations.

(82) Justification implies that the referring physician and radiologist make the decision on a radiological procedure on the basis of:

- the case history, clinical examination of the patient, and clinical laboratory results; and
- the availability of radiological examination techniques or alternative techniques.

(83) Readers are advised to refer to other publications that give examples of good radiological practice (WHO, 1983, 1987; RCR, 1993; CEC, 1996, 2000).

(84) Reiner et al. (2000) reported a substantial increase in the number of radiological examinations performed in some centres following a conversion from conventional to digital radiology. The study was performed in several US hospitals and demonstrated that the 'inpatient utilisation' (i.e. the number of examinations per inpatient day) increased by 82% (from 0.265 to 0.483 examinations per patient day) after a transition to filmless operation. This is substantially greater than the increases of 11% (from 0.190 to 0.211 examinations per patient day) seen nationally at film-based hospitals. Outpatient utilisation (i.e. the number of examinations per visit) increased by 21% (from 0.108 to 0.131 examinations per visit) compared with a net decrease of 19% (from 0.148 to 0.120 examinations per visit) nationally at film-based hospitals.

(85) These increases are probably a result of the relative ease of obtaining and archiving images with digital systems. Thus, justification criteria should be one of the key components considered in the update of a QA programme when a facility converts to digital imaging.

3.1.2. Optimisation

(86) Optimisation of digital radiology procedures (for radiation protection purposes) means the management of absorbed doses received by tissues in the region of the body being examined to a level compatible with obtaining the necessary clinical information for the particular patient (ICRP, 1991). This concept is particularly important in digital radiology because these examinations can be performed over a wide range of doses. The best images (images with least noise) are obtained with higher doses. In most cases, however, an image with acceptable diagnostic quality can be obtained with a lower radiation dose. Thus, radiological examinations should be performed with a patient dose that is as low as reasonably achievable and still commensurate with the required medical imaging purpose.

(87) There are various aspects to the optimisation of radiation protection in digital radiology. Some apply to equipment design considerations and technical methods of reducing patient dose. Others are operational approaches to reduce unnecessary patient doses by the appropriate selection of radiological examination and technical parameters. For example, with digital fluoroscopy systems, the relatively effortless production of images increases the tendency to obtain more images than clinically necessary. Thus, optimisation of radiation protection for digital radiology techniques and protocols are a critical part of the overall QA programme for such procedures.

3.2. Quality control and image management

3.2.1. Quality control

(88) Quality control is a process involving a series of distinct technical procedures provided by the imaging team that will help to ensure that the digital images from each unit are consistently of the highest quality. Image management helps to ensure that the system is maintained with minimal downtime and maximum efficiency.

(89) A radiographer, normally selected by the primary QC person, performs prescribed QC tests, usually under the guidance of a medical physicist. The medical physicist also performs tests to ensure that the digital images are optimised with respect to the radiation dose required to produce the digital images. An image management specialist should also be available to assess and correct problems, initiate repairs, and co-ordinate the maintenance of the image processing and network system.

(90) The physician-in-charge should understand the need for the programme, and take ultimate responsibility for its implementation and oversight. In summary, a comprehensive QC programme is required to ensure that image quality is optimised, that radiation doses to patients are managed properly, and that problems with the system are corrected in an effective and timely manner. Specific tasks are performed by each member of the QC team.

3.2.2. Equipment performance evaluation

(91) Before a digital radiology system can be employed for clinical use, it should pass an acceptance test. The tests conducted are designed to assess the overall imaging capability of the system and the patient dose required to achieve an acceptable clinical image. The acceptance report should show the results of each of the tests performed, including all of those specified by the manufacturer. If there are items that fail to meet the requirements or there are suggestions for improvement, these items should be corrected prior to clinical use of the equipment. Corrective actions should include the necessary repair of x-ray equipment and also include recommendations that will improve image quality (such as changes in technique factors and viewing conditions).

(92) After the initial acceptance evaluation, the unit should be surveyed annually or after repair or replacement of a major component such as the x-ray tube, AEC or digital receptors. Specific tests required for performance evaluation will vary with the type of digital system. For example, performance tests required for a full-field digital mammography system will be different from those required for a CR unit. The following list gives some of the acceptance tests recommended for most digital radiographic systems.

- Collimation assessment.
- Artefact evaluation.
- Digital receptor uniformity.

- Beam quality assessment (half-value layer measurement).
- Tube potential accuracy and reproducibility.
- Exposure reproducibility.
- Image quality.
- Monitor system gain.
- Radiation output and entrance dose.

(93) The most important measurement made by the medical physicist is the assessment of radiation dose to the patient. The dose required to achieve an acceptable clinical image is directly affected by several other parameters that are discussed in Chapter 2. Recommendations for DRLs for various types of medical radiation examinations are reviewed in Section 3.4.

(94) System gain refers to the ability of an image intensifier to produce a bright image. With digital systems, gain is one of the factors that need to be optimised in order to produce optimal image quality with an acceptable radiation dose. It is important to monitor the gain of a system because aging and improperly configured systems will result in improper values of the gain. In fluoroscopy, an improper gain can be detected through the input dose rate of the image intensifier. In digital projection radiography, each device has a particular gain that it is designed to produce. If the gain is set too high, the result is amplified electronic noise. If the gain is too low, a poor SNR results. As the SNR for an imaging system increases, the ability of the system to distinguish subtle, low-contrast lesions improves (see Section 2). It is especially important to control the gain in CR and other digital systems for consistency and optimisation.

3.2.3. Quality control tests

(95) Quality control with a digital radiographic system is required to ensure that the system performance does not degrade over time. Specific QC tests performed by a qualified and authorised person will vary with the type of equipment. A digital chest system, for example, will have a different set of QC protocols from those for a full-field digital mammography system. Typically, the protocols and procedures are specified by the manufacturer or local authority. Certain tests are scheduled to be performed at daily, weekly, monthly, or quarterly intervals. These tests are designed to ensure that the system continues to operate at the highest standards. Test results should be documented and reviewed periodically.

(96) The following list gives some of the suggested QC tests to be performed on a digital radiography system. Specific procedures and recommended frequencies will vary with the type of digital system.

- Monitor cleanliness.
- Flat-field uniformity.
- Phantom image quality.
- Contrast noise ratio.

- Repeat analysis.
- System, monitor, and filming checks.
- Cassette exposure response, uniformity, and artefacts (for a CR system).

(97) Additionally, a test image, such as the Society of Motion Picture and Television Engineers (SMPTE) test pattern (Gray et al., 1985), should be used to test the display performance of the system under conditions that simulate normal operation. The overall SMPTE image appearance should be inspected to ensure the absence of gross artefacts (e.g. blurring or bleeding of bright display areas into dark areas or aliasing of spatial resolution patterns). Display monitors used for primary interpretation should be tested at least monthly.

3.2.4. Monitoring and documentation

(98) Any facility that uses a digital imaging management system should have documented protocols for monitoring and evaluating each of the areas listed in Section 3.2. The QC team is responsible for evaluating and documenting the performance of acquisition, digitisation, compression, transmission, display, and archiving and retrieval functions of the system.

(99) A computer-trained assistant qualified to operate and maintain the data management system is also recommended, to ensure sustainable high quality and system function.

3.3. Electrical safety inspection

(100) Electrical equipment and installations need to be inspected before they are first used, after any major modifications, and subsequently at suitable intervals. Before being put into service, inspections should be carried out by those responsible for the installation and operation of the electrical equipment. These inspections need to be documented. Users are responsible for carrying out and recording regular inspections.

3.4. Diagnostic reference levels

(101) Diagnostic reference levels were introduced by the ICRP (ICRP, 1991, 1996) as a practical tool for management of patient doses in radiology. 'These levels, which are a form of investigation level, apply to an easily measured quantity, usually the absorbed dose in air or in a tissue-equivalent material at the surface of a simple standard phantom or representative patient' (ICRP, 1996). Additional advice on the application of DRLs is given in ICRP (2001b). A value lower than a DRL for a particular medical imaging task may not necessarily be an optimal value, but it indicates that the procedure utilises a patient dose within the current local practice for the procedure. A value consistently higher than a DRL may require a review to determine if the protocol for the imaging procedure needs to be altered with respect to the aspects that bear on patient dose. Example of an approach to DRLs for a

complex imaging procedure can be found in Vañó and González (2001) and Vañó et al. (2001). The results of the implementation of DRLs for common conventional (i.e. non-digital) diagnostic x-ray projections in the UK showed a 30% decrease in patient doses (Hart et al., 2002).

(102) It is of particular importance that DRL should be applied to the acquisition of digital images because of the potential for high-quality images to be produced using excessive levels of exposure. Additionally, the ease of digitally recording multiple cine runs in cardiac studies and digital photospot images in radiographic imaging makes these features susceptible to overuse.

(103) Measurement of patient entrance dose or dose rate is readily made for a digital system. Similar measurements are performed as part of a QC programme for digital equipment. They will identify equipment that has been poorly set up or calibrated, or where the performance of the image intensifier has deteriorated over a period of time. However, at this time, there is no established international consensus on the DRLs that should be applied to specific digital studies. Nevertheless, the appropriate individuals at medical imaging facilities should continue to examine the literature for pertinent information on DRLs to employ at their facility.

3.5. Auditing dose levels in digital imaging

(104) There are various approaches to auditing dose levels in digital imaging; some authors advocate the measurement of the incident or entrance surface absorbed dose for the patient ($D_{a,i}$ or $D_{a,e}$ at the patient plane), and others advocate the incident or entrance surface absorbed dose for the digital detector ($D_{a,i}$ or $D_{a,e}$ at the detector plane). There is a substantial difference between the numerical values measured and the philosophy behind the two measurement strategies.

(105) The advantage in measuring absorbed dose at the detector plane is that it can easily be incorporated into a QA programme. The actual value measured will depend on the measurement protocol used (i.e. the type and thickness of any attenuator). It can be used to compare different types of equipment and to predict the equipment on which higher doses may be anticipated. The main disadvantage is that it does not reflect the actual absorbed dose received by the patient; that is influenced by the size and composition of the patient and the clinical protocol used (Vañó et al., 2002).

(106) There will be differences in the numerical values between the two quantities (i.e. the absorbed dose at the patient plane or detector plane). These arise from the inverse square law, the attenuation in the patient, and the clinical protocol. Any difference will be particularly dependent on the tube potential and tube filtration used.

3.6. Deterministic injuries

(107) There have been several reports of deterministic injuries occurring in patients undergoing certain interventional procedures with long fluoroscopy times (Geise and O'Dea, 1999; ICRP, 2001b; Vañó et al., 1998; Wagner et al., 2000). Deterministic injuries in interventional radiology can occur with either conventional or digital

fluoroscopy as a result of a combination of factors. The main contributing factor determined in an analysis of previous incidences was found to be the use of high-dose-rate equipment (Faulkner and Vañó, 2001). Another factor was the lack of training and education in radiation management of the practitioner performing interventional procedures (Archer and Wagner, 2000). A comprehensive discussion on avoidance of radiation injuries from medical interventional procedures using fluoroscopy can be found in ICRP (2001a).

3.7. The DICOM standard

3.7.1. The DICOM standard

(108) The Digital Imaging and Communications in Medicine (DICOM) standard is the industry standard for transferral of digital radiology images and other medical information between computers (NEMA, 1996). The DICOM standard permits digital communication between medical imaging equipment and systems from various manufacturers. Such connectivity is important for cost-effectiveness in healthcare, and allows users to provide radiology services within facilities as well as for remote access. For example, workstations, digital imagers, film digitisers, shared archives, laser printers, and host computers and mainframes made by multiple manufacturers and located at one site or many sites can 'talk to one another' by means of the DICOM standard. As a result, medical images can be captured and communicated more efficiently, and therefore physicians can make diagnoses and treatment decisions more quickly. See http://medical.nema.org/ for more information.

(109) However, problems can occur if conformance to the DICOM standard is not achieved and the systems do not communicate properly. Lost or misplaced images due to non-conformance can result in repeated studies, which in turn leads to increased patient doses.

(110) In addition, organisations such as the UK Royal College of Radiologists and the American College of Radiology have produced standards for teleradiology (ACR, 1998a; RCR, 1999; SCAR, 1994) and digital imaging management (ACR, 1998b; RCR, 1999).

3.7.2. Information security issues

(111) The DICOM standard does not address security issues such as the privacy and confidentiality of all patient data. New regulations, e.g. the Health Insurance Portability and Accountability Act in the United States (HIPAA, 2000), call for strict accountability in many areas such as security, data integrity, and patient confidentiality. This requirement to ensure the privacy and quality of digital electronic images extends to intranet images within hospitals or practices as well as remote teleradiology. It may be necessary for facilities to implement multiple levels of system security, including access control and encryption. Protection of the infrastructure is important for all organisations that handle patient-identifiable healthcare information. This can be achieved by:

- preventing unauthorised access to the network or to the network assets;
- establishing access controls and authentication measures to protect access to the data; and
- ensuring data integrity and patient confidentiality (i.e. removing identifying factors from the data itself, but ensuring that the original data is protected and not compromised by changes, compression, etc.).

(112) Manufacturers of PACS are under increasing pressure to provide a high level of data security, fault tolerance, and patient privacy without compromising the major advantage of a PACS, i.e. ease of access to images and patient information. Purchasers of digital systems should be aware of these security challenges.

(113) Sometimes, if images are not easily available or transmitted to another medical facility, a procedure may be unnecessarily repeated.

4. ICRP RECOMMENDATIONS FOR DIGITAL RADIOLOGY

- **Appropriate training, particularly in the aspects of patient dose management, should be undertaken by radiologists, medical physicists, and radiographers before the clinical use of digital techniques.**
- **Local DRLs should be reviewed when new digital systems are introduced in an operational facility.**
- **Frequent patient dose audits should occur when digital techniques are introduced in an operational facility.**
- **The original image data should be made available to the user, not only for objective testing in a rigorous QA programme but also for other independent tests of the performance of digital imaging systems.**
- **When a new digital system or new postprocessing software is introduced, an optimisation programme (for radiation dose) and continuing training should be conducted in parallel.**
- **Quality control in digital radiology requires new procedures and protocols. Acceptance and constancy tests should include aspects concerning visualisation, transmission, and archiving of the images. Specialists dedicated to maintaining the network and the PACS should be available.**
- **As digital radiology images are easier to obtain and transmit in modern communication networks, referring physicians should be fully conversant with the justification criteria for requesting medical x-ray imaging procedures.**
- **Industry should promote tools to inform radiologists, radiographers, and medical physicists about the exposure parameters and the resultant patient doses. The exposure parameters and the resultant patient doses should be standardised, displayed, and recorded.**

APPENDIX A. THE FUNDAMENTALS AND ADVANTAGES OF DIGITAL SYSTEMS

Main points

- **It is characteristic for any digital radiography device that the x-ray detection stage, the final storage of the digital information, and the means for displaying the image (e.g. a workstation with adequate viewing software connected to a monitor providing diagnostic quality) are physically separate components.**
- **For the field of medical image communication, the protocols in the DICOM standard have been established. However, due to the rapid development of new technologies and methods, the compatibility and connectivity of systems from different manufacturers is still a great challenge.**
- **Storage-phosphor systems use a storage-phosphor plate to record the image from the x-ray exposure as latent information. The systems are cassette based, so existing x-ray equipment can be used without modifications. However, exposure parameters cannot automatically be stored together with the image data.**
- **The 'flat-panel' detector is the most recent development. X-ray detection and image read-out is performed in one step. An integrated assembly consisting of an x-ray converter and the read-out electronics is a common feature of these detectors. X-ray equipment with 'flat-panel' detectors is able to store exposure parameters together with the image data.**
- **One of the advantages of digitising the image information is the application of image processing to improve visualisation. As a consequence of these processing capabilities, there is no longer a strict correlation between the brightness (or blackening) of the image and the exposure.**
- **A function available in all workstations is 'windowing'; this allows specification of a subset of the full range of pixel values for display. All image information outside the window range is no longer visible. Window centre and width are usually transferred as part of the DICOM header of an image. An incorrect window setting may also be interpreted erroneously as an under- or overexposure.**
- **A poorly performing image display device can compromise the quality of the acquired data. The user should have complete information with respect to a minimum set of requirements for the image display device, either by physical specification of important display parameters or by reference to specific models of image display devices.**

A.1. Introduction

(A1) In contrast to inherently digital techniques such as CT or MRI modalities, all digital projection radiography techniques result from the continuous development of conventional (i.e. analogue) techniques. There are various different applications for digital techniques in the field of radiology, and each application has specific requirements for the x-ray equipment. The existing range of applications of conventional radiology systems can be divided into two major groups; radiography and fluoroscopy.

(A2) This subdivision can also be applied to digital x-ray equipment. However, some items of equipment combine both radiography and fluoroscopy. It should be noted that the process of development of digital systems was quite different for the two groups.

(A3) Image-intensifier-based fluoroscopy systems have been supplemented with means for intermediate image storage, i.e. LIH and hard-disk recording ('digital acquisition series'). The key component of x-ray detection in fluoroscopy, which occurs in the image intensifier television system, has hardly changed at all. Recent improvements have taken place in the television system by gradual integration of additional components for digitisation, signal processing, and recording of the image series acquired. The introduction of real-time digital image processing has improved image quality. Apart from the opportunities available due to image processing of the digital image, the basic operation of these image intensifier television systems has not changed fundamentally. The new fluoroscopy systems with flat-panel detectors allow all the functions of the digital detectors.

(A4) This is not true for direct radiography units such as Bucky tables or wall stands. The conventional x-ray detector is usually a cassette supplied with a film-screen combination. The luminescent screen acts as a first detection stage. It converts the x-ray image into light. It is the light produced by the fluorescent screen that exposes the photographic film in the second detection stage. After wet-chemical processing, the film contains the image and is the final storage device.

(A5) It is characteristic for any digital radiography device that the x-ray detection stage, final storage, and the means for displaying the image are physically separate components. This is of importance for a definitive understanding of the differences between analogue and digital radiography systems for practical applications. As a consequence, there are a few requirements that should be met in the clinical environment when digital x-ray systems are employed.

(A6) In principle, the minimum technical environment for a digital workflow consists of the x-ray device connected directly to a hardcopy printer. For enhanced usage of the digital capabilities, it is recommended that the following components, at least, should be connected within a local area network.

- A display device (e.g. a workstation with adequate viewing software) connected to a monitor providing diagnostic quality. This enables image processing to occur.
- A storage/archive device, e.g. hard disk rack, redundant array of independent disks, magneto-optical, compact disc, digital versatile disk, or at least a hardcopy printer with diagnostic quality. There is extensive information in a digital image. This, combined with the large numbers of images acquired in a digital department, places heavy demands on the storage of digital information. The storage/archive device is usually called a 'picture archiving and communication system' (PACS).

(A7) With the widespread introduction of personal computers, administrative processes in hospitals have changed significantly. Hospital information systems and radiology information systems are an integral part of an integrated data network. To benefit from such an existing data management, the digital x-ray device should also be embedded in the operations scheduling. Demographic data obtained during the admission of a patient, the name of the requesting physician, and the procedure could be directly accessible at the x-ray device. This would facilitate the workflow significantly and should reduce errors (e.g. due to duplicate data entry).

(A8) For example, for faultless, safe, and secure information exchange, a high-level standardisation process must be available, as used in telecommunication and broadcasting. For the field of medical communication, the DICOM standard has been established (Mildenberger et al., 2002). All recently introduced medical products should therefore comply with the DICOM standard. However, due to the rapid development of new technologies and methods, the compatibility and connectivity of systems from different manufacturers is still a great challenge.

A.2. Basic terminology

(A9) The main difference between analogue and digital systems from a practical point of view is the storage of the final image as a set of finite numbers. This transformation from the physical medium to a mathematical description offers an opportunity to profit from recent progress in computer technology and information processing. To maintain the quality of this processing, knowledge of selected terms is important (see Glossary).

A.3. Imaging chain of digital x-ray systems

A.3.1. X-ray conversion

(A10) The first stage in converting the information from the x-ray beam transmitted through the patient is detection of the x rays and conversion of the incoming quanta into a signal. In fluoroscopy, this is usually done by image intensifiers, a well-established technology. In an image intensifier, x rays are absorbed in a phosphor screen and converted into light photons. These photons impinge upon a photocathode that emits electrons in proportion to the number of incident x rays. These photo-electrons are accelerated across a vacuum in an image intensifier to produce an amplified light image. The light is converted by the television camera and finally digitised.

(A11) For currently available digital fluoroscopy systems, there is basically no significant difference. A brief description of future dynamic flat-panel detectors is presented in Appendix A.3.3.

(A12) For radiographic applications, the x-ray converter has changed fundamentally. Apart from the digitisation of conventional films, all modern digital detectors are based on complete replacement of the film-screen combination. Intensive research over the last 20 years has resulted in different technical solutions that have been developed and marketed.

(A13) From a physical viewpoint, detectors are often subdivided into the 'direct' and the 'indirect' conversion principle:

- direct principle: x rays are directly converted into electrical charges by a semiconductor material (e.g. amorphous selenium) without any intermediate steps; and
- indirect principle: x rays are first converted into light photons by a scintillator material (e.g. caesium iodide); the light photons are converted into electrical charges by a sensor (e.g. amorphous silicon plate with photodiodes).

(A14) From a practical viewpoint, digital detectors can also be subdivided according to their capabilities for fields of applications. Physical dimensions, image read-out speed, spatial resolution, and useful radiation dose range are the main enabling factors.

A.3.2. Storage-phosphor systems

(A15) In the meantime, the digital radiography systems used most commonly are based on storage-phosphor plates. This technology has been introduced under the designation of 'CR' systems. With the introduction of other technologies, this name is misleading, and they are named here as 'storage-phosphor systems', the term commonly used in the literature. In a storage-phosphor plate, the image from the x-ray exposure is maintained as latent information, represented by excited electrons. The image information is read by scanning the phosphor plate with a laser. The excited electrons are forced to release their energy as visible light that is detected synchronously by a sensor. Afterwards, the plate is erased with a high-intensity flashlight and can be re-used. The lifetime of storage-phosphor plates is limited by phosphor fatigue and mechanical damage such as scratches.

(A16) Storage-phosphor systems are also cassette based, so existing x-ray equipment can be used without modifications. Additionally, a stand-alone read-out device is necessary; this usually provides basic image processing features and network connectivity. Since the x-ray equipment and read-out device are not usually coupled, exposure parameters (kVp, mAs, dose–area product) cannot be stored automatically as part of the image description data (e.g. in the DICOM header).

A.3.3. Solid-state (flat-panel) detectors

(A17) The newest development is the solid-state or 'flat-panel' detector, where x-ray detection and image read-out is performed in one step, and the final image is presented a few seconds later on a display device.

(A18) The most common design is a sandwich assembly consisting of a layer with an x-ray converter deposited on an amorphous silicon plate. The amorphous silicon plate forms the image matrix with square pixels between 130 and 200 μm in size. Typically, 70% of the pixel area is associated with a photodiode that converts the light of the scintillator into an electrical charge. An electrical circuit, structured by column and row, transfers the charges collected by the pixels (i.e. by column and

Table A.1. Application and impact of detector technologies discussed in Section A.3

Applications Detector technology	Estimated impact on		
	Fluoroscopy	Radiography	Mammography
Image intensifier – television camera	+	o	–
Storage-phosphor system (CR)	–	+	o
Intensifying screen – optical camera	–	–	–
Solid-state detector CsI/a-Si ('indirect')	+	+	+
Selenium drum	–	o	–
Solid-state detector a-Se/a-Si ('direct')	o	+	+

a-Si, amorphous silicon; a-Se, amorphous selenium; CR, computed radiography; CsI, caesium iodide. Meaning of indicators: high (+), medium (o), low or none (–).

row) to amplifiers and analogue-to-digital converters in an area behind the amorphous silicon plate that is protected from radiation.

(A19) Solid-state technology has also produced dynamic detectors that provide rates up to 30 frames/s. With a useful absorbed dose range down to a few nGy per image, this class of dynamic detectors is applicable to low-dose fluoroscopy, interventional angiography, and high-resolution radiography (Table A.1).

A.4. Image processing

(A20) One of the motivations for digitising image information is the application of image processing to enhance image quality and for new medical applications (e.g. image fusion). Digital image processing has the following aspects or goals:

- image conditioning (e.g. basic corrections of the raw image dataset according to the physical properties of the device);
- image (quality) enhancement (e.g. latitude, local contrast enhancement, spatial resolution, noise suppression);
- image quantification [e.g. length and angle measurements (e.g. scoliosis, in the assessment of neurosurgery planning)]; and
- image documentation [i.e. connecting the image with demographic data of the patient, procedure information, textual and graphical annotations, exposure parameters (kV, mAs, ms, dose–area product), and reports].

(A21) Sections A.4.1–A.4.4 give a generalised description of image processing in digital x-ray systems by concentrating on common aspects. For any particular technology, there are differences in detail, and some aspects that are mentioned might not be applicable at all.

(A22) During signal processing, the image is transformed and manipulated to accomplish a certain level of fidelity with respect to the objective. In order to control and optimise the processing result, the image is available at intermediate steps. Basically, three different types of image may be distinguished:

- raw image (read-out signal of flat-panel detector or storage-phosphor system);
- original image (after all device-specific corrections); and
- processed image (for display).

A.4.1. Image conditioning

(A23) Image conditioning consists of all processing steps necessary to transform the raw image into the original image as defined above. In all types of image receptor, the physical sensitivity usually varies locally (e.g. because of the graininess of powder phosphors used in intensifying screens). In analogue x-ray systems, this leads to local variations in optical density, even if there is no local variation in x-ray exposure. In digital systems, specific algorithms can compensate for these local variations.

(A24) It should be noted that all the conditioning procedures have great impact on the final imaging performance of the digital image receptor system.

(A25) Most digital detectors show a linear gradation curve (see Fig. A.1) that does not provide the common visual impression of radiographic images experienced with conventional films. The adaptation is made by digital processing, i.e. separate from the detection process. In a digital system, the final gradation curve is a mathematical

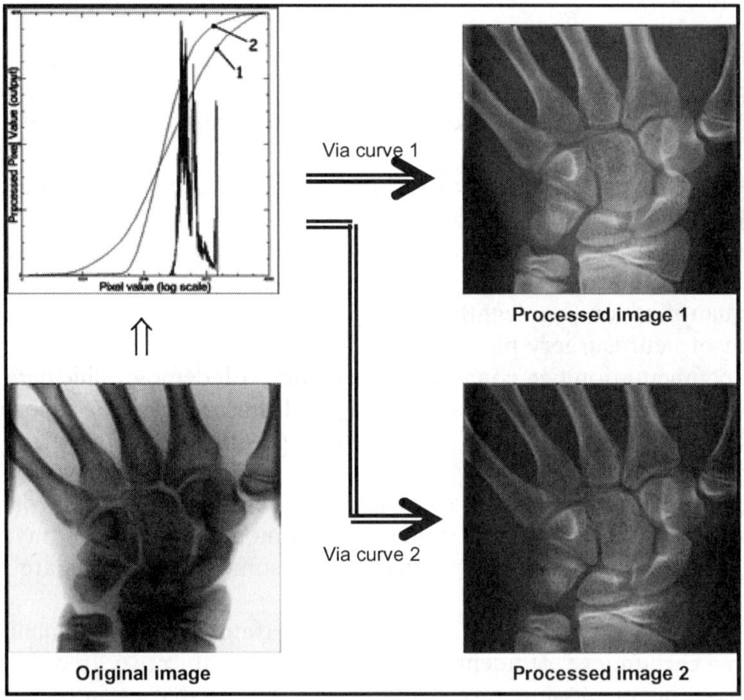

Fig. A.1. Image processing showing gradation curve, windowing, and grey-scale inversion.

calculation and not a physically fixed property as in conventional films. This provides much more flexibility because, in principle, any arbitrary shape can be generated for the gradation. Therefore, with this control of the dynamic range, contrast enhancement for selected areas of the image is possible. With specific knowledge of the body part examined, an individual optimisation is recommended and this provides further improvement of the image quality.

(A26) As a consequence of these processing capabilities, there is no longer a strict correlation between the brightness (or blackening) of the image and the radiation exposure. This means that image brightness can no longer be used as an indicator for a correct initial exposure. Therefore, it is necessary to use the recommended exposure provided in the system instructions (e.g. the 'exposure index') instead.

A.4.2. Image enhancement

Introduction

(A27) Nowadays, digital image processing algorithms provide a broad capability of image manipulation. Not all of these algorithms are applicable or useful for the field of medical imaging. All algorithms must be validated in order not to corrupt or distort the diagnostic information. Accordingly, only a subset of the algorithms is applicable in clinical practice.

(A28) Nevertheless, the final results of image processing should be checked carefully by means of QC, algorithms should be validated and controlled by a set of configuration parameters, and the processing results should be adjusted for a given medical imaging application or for a user-preferred image impression. A careful configuration should be done during the installation phase in close co-operation with the manufacturer. Incorrect configuration values may result in improper processing, resulting in the need to repeat acquisitions.

Standard processing functions

(A29) Once the pre-adjustment of image brightness and contrast has been performed using the gradation curve (i.e. image conditioning), an additional function is usually available on most radiological workstations and viewing software. This function is called 'windowing' (see Fig. A.2) and allows one to specify a subset of the full range of pixel values for a display or printout. This subset range (i.e. 'window width') is then mapped to the luminance scale of the display or hardcopy device. The position of the 'window' is defined by a second parameter called 'window centre'. Window centre and width are usually transferred as part of the DICOM header of an image.

(A30) This type of processing is used to enhance the visibility of certain details in the image. However, all image information outside the window range is no longer visible. An incorrect window setting may also be interpreted erroneously as an under- or overexposure.

(A31) In a conventional film-screen combination, the optical density is proportional to the exposure, and the photographic film shows more blackening on the more highly exposed areas. In a digital x-ray device, there is usually no intrinsic

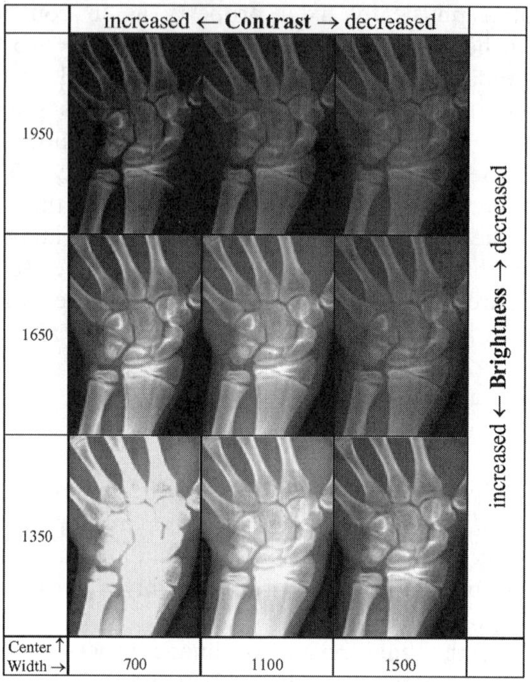

Fig. A.2. Illustration of windowing function for a 12-bit image (pixel value range from 0 to 4095). Each image is displayed with a combination of window centre and window width. Images in one row have the same centre value; images in a column have the same window width value.

inversion of the image signal. This is performed as a processing step and therefore is a selection option.

Advanced processing functions

(A32) A general problem is that the contrast of small details needs to be enhanced without overenhancing the global contrast of the whole image. This is not possible with the standard processing functions described above.

(A33) Advanced processing functions that demonstrate promising solutions for this problem are available, but most of these are manufacturer specific and an accompanying description of the implemented algorithms is not generally available. Here, the description of these functions concentrates on concepts that are used commonly. For a more detailed discussion, Beutel et al. (2000) and Stahl et al. (2000) should be consulted.

(A34) The main difference between advanced and standard processing functions is that the processing result for an individual pixel is no longer independent of the neighbouring pixels. The spatial frequency content of the image is analysed and temporarily separated. These subimages are processed differently and the results are combined to obtain the final, enhanced image. The processing scheme and the amount of enhancement is controlled by a set of parameters; these parameters need to be adjusted carefully, otherwise artefacts may appear in the final image. Some of these artefacts may be misinterpreted as pathology of a patient.

(A35) It should be noted that images with metal implants (e.g. prosthetic joints) are the most challenging examples due to the high contrast at the implant border. Incorrectly adjusted parameters may result in so-called overshoot artefacts; these can be interpreted erroneously as loosening of the implant (see Fig. A.3).

Processing functions for fluoroscopy

(A36) Digital image processing also gives opportunities for reducing patient dose in fluoroscopy. Even displaying the last image of the series if fluoroscopy has been stopped (i.e. LIH) or using graphical aids that simulate collimator blade and semi-transparent filter movement without the need for radiation significantly reduce the necessary fluoroscopy times.

(A37) In addition to the functions that are applicable for single images, digital fluoroscopy processing also takes place in the time domain (combining images acquired at different times). In the case of low-dose fluoroscopy applications, noise suppression can be performed by averaging a predefined number of preceding images. Dedicated algorithms for motion detection are used for optimising the trade off between noise suppression and blurring.

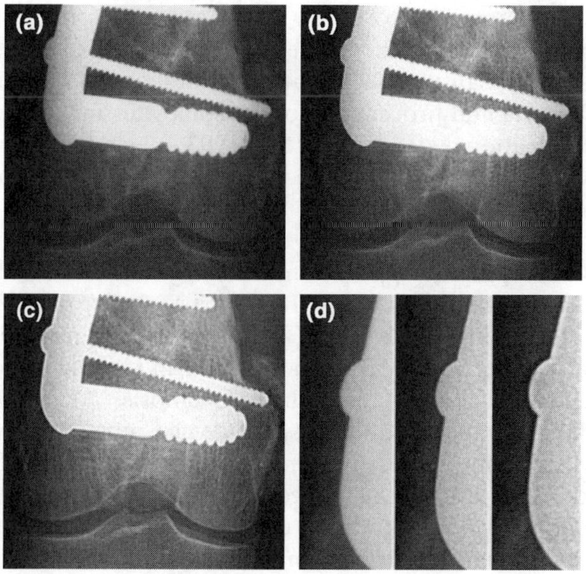

Fig. A.3. Illustration of overshoot artefacts at high-contrast implant borders. (a) Without edge enhancement, (b) kernel size 7×7 pixels, (c) kernel size 15×15 pixels, and (d) zoomed display of critical border showing black and white halos for (c).

(A38) Digital processing is also a prerequisite for pulsed fluoroscopy. It enables acquisition of x-ray images at lower frequency, but permits displaying them at a normal frame rate without flickering.

A.4.3. Image documentation (hardcopy)

(A39) Image review, documentation, and archiving may also be performed in a very similar manner as in conventional radiography (i.e. by printing on film and using a light box). A variety of different products have been available for several years, and they can be classified as either 'wet' or 'dry' systems. The technical difference is in the process of developing the printing media. 'Wet' systems expose photographic film with a laser beam scanner, and the film is developed in a wet chemical process. 'Dry' systems blacken the media via thermotransfer techniques, require less maintenance, and have only minor constraints in image quality.

(A40) In the past, hardcopy systems were connected directly to the x-ray equipment. This has recently changed to network connections following the DICOM standard. The process of printing digital images consists of three steps:

- retrieving the image data and identifying the sending modality;
- applying modality-specific processing (layout, annotation, gradation or density corrections, etc.); and
- transferring the final image information to the print media.

(A41) Each of these steps can influence the image quality of the final print result. Therefore, QC procedures need to cover the whole process from the sending modality to the display on the light box. Specific technical test images should be provided by the sending modality or workstation as well as by the hardcopy camera.

(A42) Quality criteria and procedures for constancy testing are described in detail in national and international standards (IEC, 1993, 1994a,b).

A.4.4. Image display (softcopy)

(A43) In many radiographic applications, image generation and reviewing do not take place at the same time or location. In the past, this required transporting the original film or a copy. Introduction of a PACS enables transmission of the digital images via network connection. Workstations that run viewing software are used for displaying and reviewing in order to reach the diagnosis.

(A44) Depending on the task, the display device must fulfill certain basic requirements to provide image presentation at diagnostic quality. First, room light conditions should not deteriorate image contrast by dazzling or creating flare. The display itself should provide a minimum contrast ratio, and the grey-scale presentation should be calibrated in a defined manner. Specific information is part of the DICOM standard and is described in detail as 'standardised grey-scale display function'. The goal is to maintain the same level of image quality on all connected display devices in a network environment (when used for the same purpose).

Procedures for QC are described in national and international standards (e.g. IEC, 1994b).

(A45) Finally, the image presentation by the workstation software must also be considered. Due to technical limitations, many display devices are not able to show the full image matrix generated by the X-ray detector. Image viewing software supplies zooming algorithms to handle this situation. It should be noted that dropping of pixels when fitting the image to the display area can cause aliasing artefacts. This should be suppressed by image processing (controlled spatial filtering to limit the frequency content of the displayed submatrix). Most workstations provide a specific 'full-resolution' mode, displaying exactly one image pixel by one display pixel without any interpolation ('1:1 scaling').

(A46) A poorly performing image display device can compromise the quality of the acquired data. The user should have complete information with respect to a minimum set of requirements for the image display device, either by physical specification of important display parameters or by reference to specific models of image display devices.

A.5. Image quality and imaging performance

A.5.1. Introduction

(A47) This section provides a brief introduction and some general remarks on image quality in connection with the imaging performance of a digital detector. The term 'image quality' is strictly used to describe the quality of an image under consideration.

(A48) The image quality is determined not only by the imaging performance of the detector, but also by the exposure conditions (e.g. image receptor dose, beam quality, irradiation field size), object properties (e.g. test phantoms), and image processing. Therefore, it is important to distinguish clearly between the physical 'imaging performance' of a device and the clinical 'image quality' of the final image.

A.5.2. Image quality

(A49) A summary of typical figures of merit describing the image quality is as follows:

- spatial resolution limit;
- contrast of coarse structures;
- contrast of detailed structures ('sharpness');
- noise perception ('graininess', small area signal fluctuations);
- homogeneity (large area signal fluctuations); and
- artefacts (due to defects or scratches, pixel drop-outs, etc.).

(A50) Most of these figures of merit are meaningful for conventional as well as digital systems, and a variety of technical phantoms are available. In particular, the

(a)

(b)

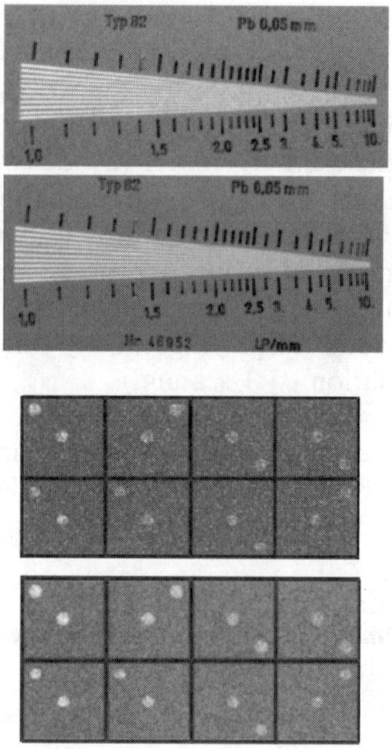

Fig. A.4. Comparison of high-contrast and low-contrast phantom. Upper row: relative dose 100%, lower row: relative dose 400% (theoretical improvement of signal-to-noise ratio by a factor of two). (a) No significant change of spatial resolution can be observed because of the high object contrast, so increased noise level does not affect the visibility. This does not correlate with clinical image quality. (b) Significant change of low-contrast resolution can be observed, which would also be expected in clinical practice.

benefit of the spatial resolution limit determined with lead bar-patterns is discussed in Section A.5.3. In the presence of aliasing, the results of lead bar-patterns do not often correlate with clinical results as real objects have a significantly lower primary contrast. More valuable results are achieved with low-contrast phantoms that better simulate practical conditions. Evaluation of low-contrast resolution covers the combination of spatial resolution and noise properties that together give a good estimate of clinical performance.

(A51) It should be noted that most results depend on exposure conditions and image processing features (see Fig. A.4). Comparable results require standardised procedures and detailed control of all relevant parameters.

A.5.3. Imaging performance

(A52) Imaging performance should be evaluated continually during clinical application and especially during the development process for the detectors. The capabilities of new detector technologies need to be compared with existing equipment in an objective, physical manner. The concept of the modulation transfer function (MTF) is used to describe how well the contrast of an object with defined size is preserved in the image (ICRU, 1986). When the MTF value is unity, the contrast has not been degraded at all. When the MTF value is zero, the detail is no longer present in the image. The MTF is usually given as a function of spatial frequency.

(A53) In systems with digital image processing, the MTF can no longer be considered a stand-alone criterion. With gradation processing, contrast enhancement by windowing, and dedicated spatial filtering of the image (as described in previous sections), contrast can be changed dramatically. On the other hand, contrast amplification will also enhance the noise perception, so in principle the detectability has not been increased at all. A more general concept is the noise perception, e.g. expressed as SNR.

(A54) In x-ray imaging, noise according to the statistical quantum fluctuation is a natural part of any image. The amount of quantum noise is determined by the total number of x-ray quanta, and can be described as the mean fluctuation of the average number of quanta. Statistical models predict the fluctuation (i.e. noise) to be proportional to the square root of the average number (i.e. signal). Calculation of the SNR due to quantum noise shows that SNR is proportional to the square root of the average number of x-ray quanta.

(A55) For a fixed beam quality, the number of x-ray quanta is directly proportional to the absorbed dose in the detector (i.e. to improve the quantum noise related to SNR by a factor of two, the absorbed dose in the detector needs to be increased by a factor of four).

(a) 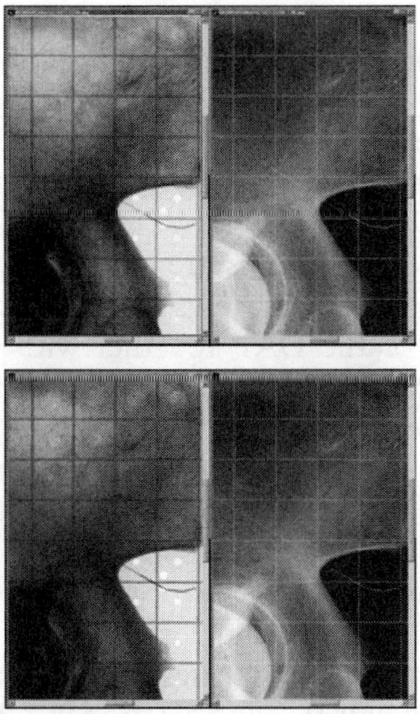

(b) 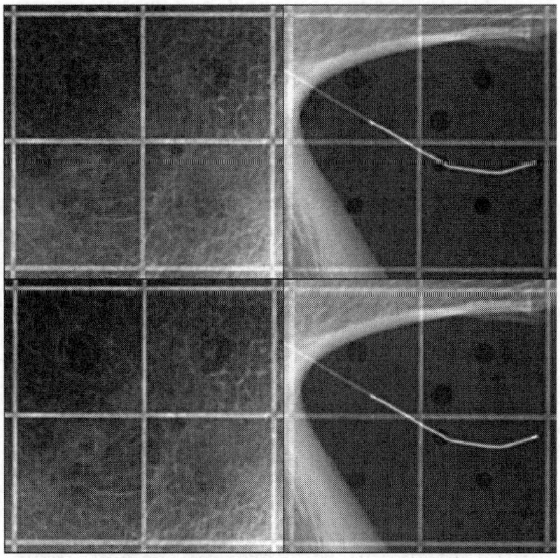

Fig. A.5. Overview and zoomed area. Original and processed image pair with different doses. (a) Examples of exposures performed with a flat-panel detector. The images in the bottom row were taken with two-fold exposure. The phantom consists of bone parts from the pelvis arranged on the CDRAD phantom (University of Nijmwegen) providing some defined low-contrast structures. A fine, high-contrast detail is simulated by the guidewire of a catheter. (b) Upper row: relative dose 100%, lower row: relative dose 200%. Only minor degradation in noise perception, even difficult to see in the printout. No degradation in spatial resolution.

(A56) An ideal x-ray detector cannot provide a better SNR in the image than that already in the radiation field due to the quantum SNR. A physical approach for a general description of imaging performance is to compare the SNR in the image to that in the radiation field. This figure of merit is called 'detective quantum efficiency' (DQE). A DQE value of unity describes an ideal detector. A DQE value of 50% means that the considered detector needs twice the dose to generate an image with the same SNR fidelity as an ideal detector. The DQE is usually given as a function of spatial frequency. The DQE of a real detector also depends on the beam quality and the absorbed dose in the detector.

(A57) The DQE is a very valuable but complex figure of merit, and measurement procedures can also affect the results significantly. Recently, an IEC working group proposed a draft standard for measuring the DQE (available at http://www.iec.ch/).

(A58) The DQE, as a function of absorbed dose in the image receptor, will provide a physically related value for the useful absorbed dose range of an x-ray detector. However, published 'sensitivity' values, following conventional film-screen categories, do not consider the issue of noise.

(A59) Fig. A.5 shows examples of pairs of images (original and processed), with each pair obtained at two different doses [Fig. A.5(a) shows the overall image (overview) and Fig. A.5(b) shows a magnified area of the image (zoomed area)].

A.6. Integrated dosimetry tools

(A60) In fluoroscopy and, especially, interventional angiography, it is common (in some countries, it is mandatory) to record the dose–area product (DAP) as a measure of the total radiation exposure to a patient. If the x-ray geometry is known, patient entrance dose at a defined reference point (see Appendix B) can be calculated additionally (ICRP, 2001b). Technically, a dosimeter chamber is integrated into the collimator of the x-ray tube.

(A61) In radiography, only a limited number of x-ray systems are supplied with such a dosimeter chamber. Additionally, only integrated systems (e.g. with flat-panel detectors) provide automatic recording and storing of the DAP together with the image data. The recorded DAP values should be part of the image description data (e.g. in the DICOM header). After completing a procedure, the DAP for all images could be accumulated and reported back (e.g. to the radiology information system or PACS). Standardisation of this data exchange is still an ongoing process and part of the DICOM activities.

(A62) As mentioned above, in digital radiography, there is no longer a fixed correlation between image brightness (or film blackening) and the absorbed dose in the image receptor. Therefore, it is recommended (in some countries, it is mandatory) that digital radiography systems provide a so-called 'exposure index'. The exposure index gives the radiographer a coarse indication of the absorbed dose in an image receptor to prevent systematic errors such as over- or underexposure.

(A63) It should be noted that the exposure index is a QC tool that provides indirect information on patient dose. It is not a replacement for the patient entrance dose or DAP.

APPENDIX B. PATIENT DOSIMETRY: QUANTITIES AND UNITS

(B1) The measurement of patient dose is regarded by many regulatory bodies as an essential feature of QA. Patient doses in a centre can be assessed and compared with DRLs. Centres with high dose levels would then be subject to optimisation studies (for radiation dose management). These studies will involve patient dosimetry measurements. It is therefore important that the measurements and dosimetry quantities used are consistent. Patient doses can be assessed by a number of methods and expressed as various quantities.

B.1. Absorbed dose

(B2) Absorbed dose (D) is the amount of energy deposited at a point in a medium ($d\bar{e}$) per unit mass (dm) (i.e. $D = d\bar{e}/dm$) (ICRU, 1993). The particular medium for which D is obtained should be specified (e.g. air, skin, or other specific tissue or organ). Absorbed dose has units of joules per kilogramme (J/kg) with the special name 'Gray' (Gy).

(B3) Absorbed dose at a point (D) is a fundamental and useful quantity for patient dosimetry.

(B4) The point where the absorbed dose is measured and the irradiation conditions should be specified, e.g.

- absorbed dose in air without backscatter (free in air) ($D_{a,i}$);
- absorbed dose in air with backscatter ($D_{a,e}$); and
- absorbed dose at a given point in tissues (with tissue scatter) ($D_{t,i}$).

where a is air, i is incident (excludes backscatter), e is entrance surface (includes backscatter), and t is soft tissue.

(B5) Other terms are often used in conventional and digital radiology such as 'entrance dose', 'incident dose', 'entrance surface dose', and 'entrance dose to the detector'. They generally refer to absorbed dose in air or some other medium at the point when the x-ray beam axis enters the patient or detector, but it is not always specified whether the quantity is measured or calculated in air or some other medium, or whether backscatter is included. This information should always be specified because differences in the numerical value of absorbed dose can be as much as 35–40%.

(B6) Since the numerical value of absorbed dose depends upon the medium in which it is expressed, the conversion factor for absorbed dose in air to absorbed dose in a different medium depends on photon energy and atomic composition of the medium. For example, the conversion factor for bone is higher than that for soft tissue or water.

B.2. Kerma

(B7) Kerma is the kinetic energy released per unit mass (units J/kg, Gy). For photons, the kinetic energy released is the kinetic energy transferred to electrons in the material. The quantity is defined with respect to the specific material in which the interactions are taking place (e.g. air kerma, water kerma, etc.).

B.3. Mean organ dose

(B8) The mean absorbed dose (D_T) in a specified tissue or organ (T) is given by:

$$D_T = 1/m_T \int m_T D \, \mathrm{d}m$$

where m_T is the mass of the tissue or organ, and D is the absorbed dose in the mass element, dm.

B.4. Equivalent dose

(B9) Equivalent dose (H_T) is given by the formula (ICRP, 1991):

$$H_T = \Sigma_R w_R D_{T,R}$$

where w_R is the radiation weighting factor, and $D_{T,R}$ is the mean absorbed dose to tissue T from type of radiation R. Photons and electrons have a w_R of 1.

B.5. Effective dose

(B10) Effective dose (E) is given by the formula (ICRP, 1991):

$$E = \Sigma_T w_T H_T$$

where w_T is the tissue weighting factor for a given tissue T (see Table B.1). Effective dose is useful in patient dosimetry studies for comparative purposes. It is inappro-

Table B.1. Values of tissue weighting factor (w_T)

Organ/tissue	w_T
Bone marrow	0.12
Lung	0.12
Bladder	0.05
Oesophagus	0.05
Bone surface	0.01
Skin	0.01
Breast	0.05
Stomach	0.12
Colon	0.12
Thyroid	0.05
Gonads	0.20
Liver	0.05
Remainder	0.05

priate to use effective dose for risk estimation in patient dosimetry as the characterisation of a given patient (e.g. age) may differ from that of a working or general population on which the values of w_T in Table B.1 are based. The unit of effective dose is J/kg, with the special name 'Sievert' (Sv).

B.6. Dose–area product

(B11) The DAP is the product of the absorbed dose in air without backscatter and the area of the x-ray field measured at the same distance from the focus, and is expressed in Gy cm^2. It can be determined at any convenient location in the beam axis between the x-ray source and the patient.

(B12) The DAP is helpful in dose control for stochastic effects to patients and operators, but is not a practical method for estimating the maximum cumulative absorbed dose to skin or useful for predicting deterministic effects.

(B13) The DAP can be measured using a large area ionisation chamber placed at the exit port of the x-ray tube, or if the x-ray unit is a modern digital system, the DAP can be calculated using data for the generator and the digitally recorded jaw positions. The DAP can be determined routinely without interfering with the imaging procedure if the appropriate monitoring devices have been installed on the x-ray equipment.

(B14) Some authors may use the quantity 'air kerma–area product' if calibration of the measurement device is for air kerma rather than absorbed dose in air.

B.7. Backscatter factor

(B15) The backscatter factor is the ratio of the entrance surface dose in air at the position corresponding to the surface of the patient, when the patient is present, to that at the same point when the patient is absent. The backscatter factor depends upon the x-ray field size and the x-ray energy spectrum (i.e. usually specified by the half-value layer, x-ray tube potential, and x-ray tube filtration). Typical backscatter factors are in the range of 1.2–1.4.

B.8. Skin dose

(B16) Skin dose is the mean absorbed dose to a specific area of the skin and it is an important quantity in the evaluation of radiographic examinations or for interventional procedures where there is a risk of deterministic effects (ICRP, 2001b). The maximum cumulative skin dose (or 'peak skin dose') [i.e. the highest absorbed dose (including backscatter) received by a specific area of the skin] is the quantity that should be assessed if there is a risk of deterministic injuries.

APPENDIX C. OUTLINE FOR GENERAL EDUCATION AND TRAINING ASPECTS IN DIGITAL RADIOLOGY

The following outline was proposed by the Euopean DIMOND Consortium (Peer et al., 2003).

- X-ray systems for digital imaging
  - Capability to transfer the radiographic (kV, mAs) and geometrical parameters (distances, filed size, x-ray beam orientation, etc.) in digital format.
  - Digital detector to capture the image and its properties.
  - Sensitivity range.
  - Fluoroscopy: operation of continuous and pulsed fluoroscopy modes, LIH, high filtration, different image intensifier sizes, different fluoroscopy modes, semitransparent filters, etc.
- Image acquisition
  - Process of image acquisition.
  - Exposure level (speed class equivalent).
  - Fluoroscopy: influence of variations in the distance from object image to intensifier on absorbed dose and image quality, influence of magnification modes on absorbed dose, different fluoroscopy and image acquisition modes, etc.
- Image post processing
  - Techniques related to post processing and their influence on the image: edge enhancement, unsharp masking, dynamic range reduction, multiscale processing, noise reduction, window/level setting, look up table, etc.
  - Possibilities of image manipulation tools such as shuttering, magnification, electronic collimation, etc.
  - Fluoroscopy: image subtraction, pixel shifting, remasking, road mapping, etc.
- Image transfer
  - Process of image identification.
    Error sources in identification and their consequences in a PACS.
  - Transfer speed.
- Image analysis
  - Observable differences in different imaging modalities.
  - Possibilities of further post processing on a diagnostic workstation.
  - Possible artefacts in digital images.
  - Fluoroscopy: application of cine mode, difference in subtracted and not subtracted images, etc.
- QC
  - Recommended QC and frequency.
  - X-ray system.
  - Viewing monitors.
  - Postprocessing tools.
  - Transmission and archiving images.

- Security aspects.
- Incidents and their consequences (interruption of transmission in the network, fault in electrical power supply, etc.).

• Aspects in digital visualisation and reporting

• Workstations
  - Differences between workstations for reporting, viewing, and post processing.

• Image presentation
  - Different display modes and their usefulness for softcopy reporting.
  - Image resolution and display resolution, and influence of magnification on display resolution.
  - Different compression levels.

• Image transfer and archiving
  - Differences between lossy and lossless image compression.
  - Workflow concepts in digital environments (HIS-RIS-PACS).
  - Differences in short- and long-term archiving.
  - Legal issues concerning electronic archiving of medical images.
  - Methods available for image transfer in teleradiology.
  - Legal issues concerning teleradiology applications.
  - Legal and safety issues relevant for PACS and teleradiology.

• Patient dose management
  - Relationship between image quality and patient dose.
  - Level of image quality needed for a clinical task.
  - Actions that can influence image quality and patient dose.
  - Optimisation (of radiation dose) in digital radiology.
  - Displaying and taking advantage of dose-related parameters.
  - Patient dosimetry and DRLs.

5. REFERENCES

ACR Standard for Teleradiology (Res 21) (1994). The American College of Radiology, Revised 1996 (Res. 26) and 1998a (Res. 35). Effective 1/1/99 ACR, Annapolis Junction, MD.

ACR Standard for Digital Image Data Management (Res. 15) (1998b). The American College of Radiology. Effective 1/1/99 ACR, Annapolis Junction, MD.

Archer, B.R., Wagner, L.K. (2000). Protecting patients by training physicians in fluoroscopic radiation management. J. Appl. Clin. Med. Phys. 1, 32–37.

Axelsson, B., Boden, K., Fransson, S.G., Hansson, I.B., Persliden, J., Witt, H.H. (2000). A comparison of analogue and digital techniques in upper gastrointestinal examinations: absorbed dose and diagnostic quality of the images. Eur. Radiol. 10, 1351–1354.

Bergh, B., Felix, R. (1999). WWW-based access to radiological patient data: two years of experience. Stud. Health Technol. Inform. 64, 217–229.

Bergh, B., Pietsch, M., Vogl, T.J. (2000). Internet technology for clinical applications in a digital radiography department. Radiologe 40, 371–387.

Beutel, J., Kundel, H., Van Metter, R. (Eds.), 2000. Handbook of Medical Imaging, Volumes I, II and III. SPIE – The International Society for Optical Engineering, Bellingham.

Busch, H.P. (1997). Digital radiography for clinical applications. Eur. Radiol. 7, 66–72.

Busch, H.P. (2003). Management of dose and quality in digital radiography. Fortschr. Röntgenstr. 175, 17–19.

Busch, H.P., Busch, S., Decker, C., Schilz, C. (2003). Image quality and exposure dose in digital projection radiography. Fortschr. Röntgenstr. 175, 32–37.

CEC (1996). European Guidelines on Quality Criteria for Diagnostic Radiographic Images. EUR 16260. Commission of the European Communities, Luxembourg.

CEC (2000). Referral Criteria for Imaging Radiation. Radiation Protection 118. Adapted by experts representing European Radiology and Nuclear Medicine. In conjunction with the UK Royal College of Radiologists. Co-ordinated by the European Commission, Directorate General for the Environment, Luxembourg. Available at: http://europa.eu.int/comm/energy/nuclear/radioprotection/publication/118_en.htm.

Faulkner, K. (2001). Dose displays and record keeping. Radiat. Prot. Dosimetry 94, 143–145.

Faulkner, K., Vañó, E. (2001). Deterministic effects in interventional radiology. Radiat. Prot. Dosimetry 94, 95–98.

FDA (2001). Technical Electronic Product Radiation Safety Standards Committee, 28th Meeting. Food and Drug Administration, Department of Health and Human Services, Center for Devices and Radiological Health, Rockville, MD.

Fink, C., Hallscheidt, P.J., Noeldge, G., et al. (2002). Clinical comparative study with a large-area amorphous silicon flat-panel detector: image quality and visibility of anatomic structures on chest radiography. AJR Am. J. Roentgenol. 178, 481–486.

Geise, R.A., O'Dea, T.J. (1999). Radiation dose in interventional fluoroscopic procedures. Appl. Radiat. Isot. 50, 173–184.

Gray, J.E., Lisk, K.G., Haddick, D.H., et al. (1985). SMPTE test pattern for video displays and hard copy cameras. Radiology 154, 519–527.

Hart, D., Hillier, M.C., Wall, B.F. (2002). Doses to Patients from Medical X-ray Examinations in the UK – 2000 Review. NRPB W14. National Radiological Protection Board, Chilton, UK.

Hennigs, S.P., Garmer, M., Jaeger, H.J., et al. (2001). Digital chest radiography with a large-area flat-panel silicon X-ray detector: clinical comparison with conventional radiography. Eur. Radiol. 11, 1688–1696.

Herrmann, A., Bonel, H., Stabler, A., et al. (2002). Chest imaging with flat-panel detector at low and standard doses: comparison with storage phosphor technology in normal patients. Eur. Radiol. 12, 385–390.

HIPAA (2000). Health Insurance Portability and Accountability Act of 1996. US Public Law 104–191, 104th Congress. Final Rule published in the Federal Register (65 FR 82462) US Dept of Health and Human Services, Washington, D.C.

Huda, W., Rill, L.N., Bruner, A.P. (1997). Relative speeds of Kodak computed radiography phosphors and screen-film systems. Med. Phys. 24, 1621–1628.
IEC/TR2 61223-1 (1993). Ed. 1.0. Evaluation and Routine Testing in Medical Imaging Departments – Part 1: General Aspects. http://www.iec.ch.
IEC 61223-2-4 (1994a). Ed. 1.0. Evaluation and Routine Testing in Medical Imaging Departments – Part 2–4: Constancy Tests – Hard Copy Cameras. http://www.iec.ch.
IEC 61223-2-5 (1994b). Ed. 1.0. Evaluation and Routine Testing in Medical Imaging Departments – Part 2–5: Constancy Tests – Image Display Devices. http://www.iec.ch/.
IAEA (1996). Basic Safety Standards. International Atomic Energy Agency, Vienna.
ICRP (1977). Recommendations of the International Commission on Radiological Protection. ICRP Publication 26, Ann. ICRP.
ICRP (1991). 1990 Recommendations of the International Commission on Radiological Protection. ICRP Publication 60, Ann. ICRP 21(1–3).
ICRP (1996). Radiological protection and safety in medicine. ICRP Publication 73, Ann. ICRP 26(2).
ICRP (2001a). Radiation and your patient: a guide for medical practitioners. Also includes: Diagnostic reference levels in medical imaging – review and additional advice. Supporting Guidance 2, Ann. ICRP 31(4).
ICRP (2001b). Avoidance of radiation injuries from medical interventional procedures. ICRP Publication 85, Ann. ICRP 30(2).
ICRU (1986). Modulation Transfer Function of Screen-film Systems. ICRU Report 41. Nuclear Technology Publishing, UK.
ICRU (1993). Quantities and Units in Radiation Protection Dosimetry. ICRU Report 51. Nuclear Technology Publishing, UK.
Kunz, G., Busch, H.P. (2000). Examination of the image quality from digital fluoroscopy equipment with 1k- and 2k-image matrix. Rofo. Fortschr. Geb. Rontgenstr. Neuen. Bildgeb. Verfahr. 172, 707–713.
Launders, J.H., Cowen, A.R., Bury, R.F., Hawkridge, P. (2001). Towards image quality, beam energy and effective dose optimisation in digital thoracic radiography. Eur. Radiol. 11, 870–875.
Mildenberger, P., Eichelberg, M., Martin, E. (2002). Introduction to the DICOM standard. Eur. Radiol. 12, 920–927.
Mosser, H.M., Pärtan, G., Maltsidis, A., Hruby, W. (2002). Integration of healthcare information: quality assurance of exam optimisation with automatic transfer of exposure dose parameters into the electronic patient record. European Congress of Radiology, B-0501 ECR, Vienna, Austria.
NEMA (1996). Digital Imaging and Communications in Medicine (DICOM). National Electrical Manufacturers' Association, Rosslyn, VA, pp. 3.1-1996–3.13-1996.
Partan, G., Mayrhofer, R., Mosser, H., Mahdi, T., Hruby, W. (2001). Radiation dose of digital vs conventional fluororadiography of the upper GI tract. Eur. Radiol. 11, 2109–2111.
Peer, S., Faulkner, K., Torbica, P., et al. (2003). Relevant Training Issues for Introduction of Digital Radiology: Results of a Survey. DIMOND Report. http://www.dimond3.org/index_2.htm.
Peters, S.E., Brennan, P.C. (2002). Digital radiography: are the manufacturers' settings too high. Optimisation of the Kodak digital radiography system with aid of the computed radiography dose index. Eur. Radiol. 12, 2381–2387.
Reiner, B.I., Siegel, E.L., Flagle, C., Hooper, F.J., Cox, R.E., Scanlon, M. (2000). Effect of filmless imaging on the utilization of radiologic services. Radiology 215, 163–167.
Rill, L.N., Brateman, L., Arreola, M. (2003). Evaluating radiographic parameters for mobile chest computed radiography: phantoms, image quality and effective dose. Med. Phys. 30, 2727–2735.
RCR (1993). Making the Best Use of a Department of Radiology. Royal College of Radiologists, London.
RCR (1999). Guide to Information Technology in Radiology: Teleradiology and PACS. Board of Faculty of Clinical Radiology, Royal College of Radiologists, London.
Samei, E., Hill, J.G., Frey, G.D., Southgate, W.M., Mah, E., Delong, D. (2003). Evaluation of a flat panel digital radiographic system for low-dose portable imaging of neonates. Med. Phys. 30, 601–607.
Slone, R.M., Muka, E., Pilgram, T.K. (2003). Irreversible JPEG compression of digital chest radiographs for primary interpretation: assessment of visually lossless threshold. Radiology 228, 425–429.

Smith, J.J., Berlin, L. (2001). Picture archiving and communication systems (PACS) and the loss of patient examination records. AJR Am. J. Roentgenol. 176, 1381–1384.

SCAR (1994). Understanding Teleradiology. Technology Forum. Society for Computer Applications in Radiology. http://www.scarnet.org/.

Stahl, M., Aach, T., Dippel, S. (2000). Digital radiography enhancement by nonlinear multiscale processing. Med. Phys. 27, 56–65.

Strotzer, M., Gmeinwieser, J., Volk, M., et al. (1998). Clinical application of a flat-panel X-ray detector based on amorphous silicon technology: image quality and potential for radiation dose reduction in skeletal radiography. AJR Am. J. Roentgenol. 171, 23–27.

Strotzer, M., Volk, M., Frund, R., Hamer, O., Zorger, N., Feuerbach, S. (2002). Routine chest radiography using a flat-panel detector: image quality at standard detector dose and 33% dose reduction. AJR Am. J. Roentgenol. 178, 169–171.

Tuinenburg, J.C., Koning, G., Hekking, E., et al. (2000). American College of Cardiology/European Society of Cardiology international study of angiographic data compression phase II. The effects of varying JPEG data compression levels on the quantitative assessment of the degree of stenosis in digital coronary angiography. Eur. Heart J. 21, 679–686.

Vañó, E., Arranz, L., Sastre, J.M., et al. (1998). Dosimetric and radiation protection considerations based on some cases of patient skin injuries in interventional cardiology. Br. J. Radiol. 71, 510–516.

Vañó, E., González, L. (2001). Approaches to establishing reference levels in interventional radiology. Radiat. Prot. Dosimetry 94, 109–112.

Vañó, E., Faulkner, K., Padovani, R., Zoetelief, J., Toivonen, M., Neofostistou, V. (2001). DIMOND. A European approach to establish and use reference levels in fluoroscopy-guided procedures. *Int. Conf. on the Radiol. Protec. of Patients*, Malaga, 26–30 March 2001, IAEA-CN-85-194, pp. 444–448.

Vañó, E., Fernández, J.M., Ten, J.I., Guibelalde, E., Gonzalez, L., Pedrosa, C.S. (2002). Real-time measurement and audit of radiation dose to patients undergoing computed radiography. Radiology 225, 283–288.

Wagner, L.K., Archer, B.R., Cohen, A.M. (2000). Management of patient skin dose in fluoroscopically guided interventional procedures. J. Vasc. Interv. Radiol. 11, 25–33.

Weatherburn, G.C., Bryan, S. (1999). The effect of a picture archiving and communication system (PACS) on patient radiation doses for examination of the lateral lumbar spine. Br. J. Radiol. 72, 534–545.

WHO (1983). Effective Choices for Diagnostic Imaging in Clinical Practice. WHO Technical Report 898. World Health Organisation, Geneva.

WHO (1987). Rational Use of Diagnostic Imaging in Paediatrics. Technical Report 757. World Health Organisation, Geneva.

Zoetelief, J., Faulkner, K. (2001). Equipment requirements and specification for digital and interventional radiology. Radiat. Prot. Dosimetry 94, 43–48.

The ICRP Database of Dose Coefficients: Workers and Members of the Public
Version 2.01

Edited by

ICRP (prepared by the Task Group on Dose Calculations of Committee 2 of the International Commission on Radiological Protection)

Description

Over the last few years *ICRP* has developed a number of reports giving ingestion and inhalation dose coefficients for members of the public. A summary report, *ICRP Publication 72*, was issued giving a compilation of effective dose coefficients for intakes by inhalation and ingestion for nearly 800 radionuclides of 91 elements. Earlier, in 1994 a similarly wide-ranging set of dose coefficients for workers, based on the same biokinetic models, had been issued in *ICRP Publication 68*. Both of these Publications give only committed effective doses and in the case of inhalation cover only 1μm AMAD aerosols for the public, and 1μm and 5μm AMAD aerosols for workers.

A CD-ROM is now available which gives inhalation dose coefficients for ten aerosol sizes (0.001, 0.003, 0.01, 0.03, 0.1, 0.3, 1, 3, 5, and 10 μm AMAD) as well as ingestion coefficients. Effective doses and equivalent doses for all important tissues for a range of integration times (1, 7, and 30 days, 1, 5, 10, 20, 30, and 45 years) are given, together with the dose coefficients to age 70 years.

The CD-ROM also contains substantial help files, these contain: text from ICRP *Publication 68*; text from *ICRP Publication 72*; further information specific to the database; tables and diagrams describing the biokinetic models used for all elements.

The CD-ROM will run under Windows 98, Me, NT, XP, or 2000. It is possible to load all retrieved data directly into Excel (97 or later) by using the Export button on the Results screen.

This CD-ROM can be ordered from Elsevier Science (E-mail, USA: usinfo-f@elsevier.com, rest of the world: nlinfo-f@elsevier.nl).

- ISBN: 0-08-043876-8, Year 2001
- Additional copies at Eur 50/US$ 50 each with the same order

Price: Order form

- EUR 139.50
- GDP 93

The ICRP Database of Dose Coefficients: Embryo and Fetus (ICRP CD-ROM 2)
Version 1

Edited by

ICRP (prepared by the Task Group on Dose Calculations of Committee 2 of the International Commission on Radiological Protection)

Description

In its *Publications 56, 67, 69, 71*, and *72*, ICRP provided age-specific biokinetic models and dose coefficients (dose per unit intake of radioactive substance) for members of the public. Committed effective doses for inhalation or ingestion of radionuclides by occupationally exposed workers were given in *Publication 68*. A recent complementary report (*Publication 88*) addresses doses to the embryo/fetus after intakes by the mother before or during pregnancy. Selected radionuclides of 31 elements are considered. New biokinetic and dosimetric models for doses to the embryo/fetus are developed; they take account of transfer of radionuclides across the placenta, distribution and retention of radionuclides in fetal tissues, growth of the fetus, and photon irradiation from radionuclides in the placenta and maternal tissues. Human and animal data are used as available.

A CD-ROM is now available which gives inhalation dose coefficients for ten aerosol sizes (0.001, 0.003, 0.01, 0.03, 0.1, 0.3, 1, 3, 5 and 10 um AMAD) as well as ingestion coefficients. Effective doses and equivalent doses for all important tissues for a range of post natal integration times (1, 5, 10 and 20 years) are given, together with the dose coefficients to age 70 years. In utero and post-natal doses are given, together with the total dose to offspring. Various intake scenarios are considered as in *Publication 88.*

The CD-ROM also contains substantial help files, these contain the text from ICRP *Publication 88*, and further information specific to the database.

The CD-ROM will run under Windows 98, Me, NT, XP, or 2000.

Contents

1. GENERAL INFORMATION. Database information. Scope of the database. Radionuclides. Workers and adult members of the public. Age groups. Selecting the appropriate f_1 value or lung absorption type. Aerosol sizes. Results. Target tissues. Printing or exporting results. Data files. **Application of equivalent dose. Lung deposition fractions.** Adult female member of the public. Adult female worker.

2. TEXT FROM ICRP PUBLICATION 88. Editorial. Preface. Glossary of terms (through the Search facility). Introduction. Background. The present position. Model development. Maternal intakes. Dose coefficients. **Development of the embryo and fetus.** Introduction. Fertilisation to gastrulation. Formation of tissues and organ systems. Placenta, extra-embryonic membranes, and yolk sac. Modelling fetal organ development. Newborn child. **Biokinetic and dosimetric models.** Introduction. Maternal intakes of radionuclides. Transfer of radionuclides to the embryo, fetus, and newborn child. Development of models. Dosimetry. **Biokinetic data and models.** Introduction. Data for selected radionuclides of 31 elements. **References. Annex A Transfer of alkaline earth elements to the fetus. Annex B Iodine biokinetics in the mother and fetus. Annex C Decay products requiring C_F:C_M ratios and fetal distribution data.**

3. DATABASE LOOKUP PROGRAM: DOSE COEFFICIENTS. Exposure. Public. Occupational. **Intake.** 31 important radionuclides. **Intake scenario(s).** Acute intake, 8 different times before and after conception. Chronic intake, 3 different time frames in relation to conception. **Intake route(s).** Ingestion and/or inhalation. Ten aerosol sizes. **Output.** All tissues or a customised list. Five integration periods or a customised list. Results can be saved to a file for further processing. Lists of important nuclides can be saved for future use.

Bibliographic and Ordering Information

For information about conditions of sale, ordering procedures, and links to our regional sales offices, please read through our ordering information.

Year 2002
CD-ROM

- ISBN: 0-08-044188-2
- Windows 95, 98, Me, 2000, or XP
- Price: Order form
    - USD 149
    - EUR 149

        - Additional copies at EUR 50.00 / US$ 50.00 each with the same order